D0820599

ETHICS AND INTEGRITY IN HEALTH AND LIFE SCIENCES RESEARCH

ADVANCES IN RESEARCH ETHICS AND INTEGRITY

Series Editor: Dr Ron Iphofen *FAcSS, Independent Consultant, France*

Recent Volumes:

Volume 1: FINDING COMMON GROUND: Consensus in Research Ethics Across the Social Sciences. *Edited by Ron Iphofen*

Volume 2: THE ETHICS OF ONLINE RESEARCH. *Edited by Kandy Woodfield*

Volume 3: VIRTUE ETHICS IN THE CONDUCT AND GOVERNANCE OF SOCIAL SCIENCE RESEARCH. *Edited by Nathan Emmerich*

SERIES EDITORIAL ADVISORY GROUP

ADVANCES IN RESEARCH ETHICS AND INTEGRITY
VOLUME 4

ETHICS AND INTEGRITY IN HEALTH AND LIFE SCIENCES RESEARCH

EDITED BY

ZVONIMIR KOPORC
Catholic University of Croatia, Zagreb

United Kingdom – North America – Japan
India – Malaysia – China

Emerald Publishing Limited
Howard House, Wagon Lane, Bingley BD16 1WA, UK

First edition 2019

Reprints and permissions service
Contact: permissions@emeraldinsight.com

British Library Cataloguing in Publication Data
A catalogue record for this book is available from the British Library

ISBN: 978-1-78743-572-8 (Print)
ISBN: 978-1-78743-571-1 (Online)
ISBN: 978-1-78743-918-4 (Epub)

ISSN: 2398-6018 (Series)

ISOQAR certified Management System, awarded to Emerald for adherence to Environmental standard ISO 14001:2004.

Certificate Number 1985
ISO 14001

INVESTOR IN PEOPLE

CONTENTS

Series Preface — *ix*

About the Series Editor — *xiii*

About the Editor — *xv*

List of Contributors — *xvii*

Introduction: Research Production in Life Sciences
Zvonimir Koporc — *1*

Promoting Equity and Preventing Exploitation in International Research: The Aims, Work, and Output of the TRUST Project
Julie Cook, Kate Chatfield and Doris Schroeder — *11*

Ebola Virus Disease: A Lesson in Science and Ethics
Nicola Petrosillo and Rok Čivljak — *33*

Ethics Challenges in the Digital Era: Focus on Medical Research
Albena Kuyumdzhieva — *45*

Big Data in Healthcare and the Life Sciences
Janet Mifsud and Cristina Gavrilovici — *63*

Shaping a *Culture of Safety and Security* in Research on Emerging Technologies: Time to Move beyond "Simple Compliance" Ethics
Monique Ischi and Johannes Rath — *85*

Governing Gene Editing in the European Union: Legal and Ethical Considerations
Mihalis Kritikos — *99*

ARRIGE: Toward a Responsible Use of Genome Editing
François Hirsch and Lluis Montoliu — *115*

Dual Use in Neuroscientific and Neurotechnological Research: A Need for Ethical Address and Guidance
James Giordano and Kathinka Evers 129

Ethical Challenges of Informed Consent, Decision-Making Capacity, and Vulnerability in Clinical Dementia Research
Pablo Hernández-Marrero, Sandra Martins Pereira, Joana Araújo and Ana Sofia Carvalho 147

Diet Therapy Effective Treatment but also Ethical and Moral Responsibility
Jasenka Gajdoš Kljusurić 169

The Mismatch of Nutrition and "Medical Practice": The Wayward Science of Nutrition in Human Health
T. Colin Campbell and T. Nelson Campbell 185

Index 203

SERIES PREFACE

This book series, *Advances in Research Ethics and Integrity*, grew out of foundational work with a group of Fellows of the UK Academy of Social Sciences (AcSS) who were all concerned to ensure that lessons learned from previous work were built upon and improved in the interests of the production of robust research practices of high quality. Duplication or unnecessary repetitions of earlier research and ignorance of existing work were seen as hindrances to research progress. Individual researchers, research professions, and society all suffer in having to pay the costs in time, energy, and money of delayed progress and superfluous repetitions. There is little excuse for failure to build on existing knowledge and practice given modern search technologies unless selfish "domain protectionism" leads researchers to ignore existing work and seek credit for innovations already accomplished. Our concern was to aid well-motivated researchers to quickly discover existing progress made in ethical research in terms of topic, method, and/or discipline and to move on with their own work more productively and to discover the best, most effective means to disseminate their own findings so that other researchers could, in turn, contribute to research progress.

It is true that there is a plethora of ethics codes and guidelines with researchers left to themselves to judge those more appropriate to their proposed activity. The same questions are repeatedly asked on discussion forums about how to proceed when similar long-standing problems in the field are being confronted afresh by novice researchers. Researchers and members of ethics review boards alike are faced with selecting the most appropriate codes or guidelines for their current purpose, eliding differences and similarities in a labyrinth of uncertainty. It is no wonder that novice researchers can despair in their search for guidance and experienced researchers may be tempted by the "checklist mentality" that appears to characterize a meeting of formalized ethics "requirements" and permit their conscience-free pursuit of a cherished program of research.

If risks of harm to the public and to researchers are to be kept to a minimum and if professional standards in the conduct of scientific research are to be maintained, the more that fundamental understandings of ethical behavior in research are shared the better. If progress is made in one sphere, all gain from it being generally acknowledged and understood. If foundational work is conducted, all gain from being able to build on and develop further that work.

Nor can it be assumed that formal ethics review committees are able to resolve the dilemmas or meet the challenges involved. Enough has been written about such review bodies to make their limitations clear. Crucially they cannot follow researchers into the field to monitor their every action; they cannot anticipate all of the emergent ethical dilemmas nor, even, follow through to the publication of findings. There is no adequate penalty for neglect through incompetence, nor worse, for conscious omissions of evidence. We have to rely upon the "virtues" of the individual researcher alongside the skills of journal and grant reviewers. We need constantly to monitor scientific integrity at the corporate and at the individual level. These are issues of "quality" as well as morality.

Within the research ethics field new problems, issues, and concerns and new ways of collecting data continue to emerge regularly. This should not be surprising as social, economic, and technological change necessitate constant re-evaluation of research conduct. Standard approaches to research ethics such as valid informed consent, inclusion/exclusion criteria, vulnerable subjects, and covert studies need to be reconsidered as developing social contexts and methodological innovation, interdisciplinary research, and economic pressures pose new challenges to convention. Innovations in technology and method challenge our understanding of "the public" and "the private". Researchers need to think even more clearly about the balance of harm and benefit to their subjects, to themselves, and to society. This series proposes to address such new and continuing challenges for both ethics committees and researchers in the field as they emerge. The concerns and interests are global and well recognized by researchers and commissioners alike around the world but with varying commitments at both the "procedural" and the "practical" levels. This series is designed to suggest realistic solutions to these challenges – this "practical" angle is the USP for the series. Each volume will raise and address the key issues in the debates, but also strive to suggest ways forward that maintain the key ethical concerns of respect for human rights and dignity, while sustaining pragmatic guidance for future research developments. A series such as this aims to offer practical help and guidance in actual research engagements as well as meeting the often varied and challenging demands of research ethics review. The approach will not be one of abstract moral philosophy; instead, it will seek to help researchers think through the potential harms and benefits of their work in the proposal stage and assist their reflection of the big ethical moments that they face in the field often when there may be no one to advise them in terms of their societal impact and acceptance.

While the research community can be highly imaginative both in the fields of study and methodological innovation, the structures of management and funding, and the pressure to publish to fulfill league table quotas can pressure researchers into errors of judgment that have personal and professional consequences. The series aims to adopt an approach that promotes good practice and sets principles, values, and standards that serve as models to aid successful research outcomes. There is clear international appeal as commissioners and researchers alike share a vested interest in the global promotion of professional virtues that lead to the public acceptability of good research. In an increasingly global world in research terms, there is little point in applying too localized a morality, nor one that implies a solely Western hegemony of values. If standards "matter," it seems evident that they should "matter" to and for all. Only then can the growth of interdisciplinary and multinational projects be accomplished effectively and with a shared concern for potential harms and benefits. While a diversity of experience and local interests is acknowledged, there are existing, proven models of good practice which can help research practitioners in emergent nations build their policies and processes to suit their own circumstances. We need to see that consensus positions effectively guide the work of scientists across the globe and secure minimal participant harm and maximum societal benefit – and, additionally, that instances of fraudulence, corruption, and dishonesty in science decrease as a consequence.

Perhaps some forms of truly independent formal ethics scrutiny can help maintain the integrity of research professions in an era of enhanced concerns over data security, privacy, and human rights legislation. But it is essential to guard against rigid conformity to what can become administrative procedures. The consistency we seek to assist researchers in understanding what constitutes "proper behavior" does not imply uniformity. Having principles does not lead inexorably to an adherence to principlism. Indeed, sincerely held principles can be in conflict in differing contexts. No one practice is necessarily the best approach in all circumstances. But if researchers are aware of the range of possible ways in which their work can be accomplished ethically and with integrity, they can be free to apply the approach that works or is necessary in their setting. Guides to "good" ways of doing things should not be taken as the "only" way of proceeding. A rigidity in outlook does no favors to methodological innovation, nor to the research subjects or participants that they are supposed to "protect". If there were to be any principles that should be rigidly adhered to they should include flexibility, open-mindedness, the recognition of the range of challenging situations to be met in the field – principles that in essence amount to a sense of proportionality. And these principles should apply equally to researchers and ethics reviewers alike. To accomplish that requires ethics reviewers to think afresh about each new research proposal, to detach from pre-formed opinions and prejudices, while still learning from and applying the lessons of the past. Principles such as these must also apply to funding and commissioning agencies, to research institutions, and to professional associations and their learned societies. Our integrity as researchers demands that we recognize that the rights of our funders and research participants and/or "subjects" are to be valued alongside our cherished research goals and seek to embody such principles in the research process from the outset. This series will strive to seek just how that might be accomplished in the best interests of all.

By
Ron Iphofen (Series Editor)

ABOUT THE SERIES EDITOR

Ron Iphofen, FAcSS, is Executive Editor of the Emerald book series *Advances in Research Ethics and Integrity* and edited Volume 1 in the series, *Finding Common Ground: Consensus in Research Ethics Across the Social Sciences* (2017). He is an Independent Research Consultant, a Fellow of the UK Academy of Social Sciences, the Higher Education Academy, and the Royal Society of Medicine. Since retiring as Director of Postgraduate Studies in the School of Healthcare Sciences, Bangor University, his major activity has been as an adviser to the European Commission (EC) and its agencies, the European Research Council (ERC), and the Research Executive Agency (REA) on both the Seventh Framework Programme (FP7) and Horizon 2020. His consultancy work has covered a range of research agencies (in government and independent) across Europe. He was Vice Chair of the UK Social Research Association, updated their Ethics Guidelines and now convenes the SRA's Research Ethics Forum. He was scientific consultant on the EC RESPECT project – establishing pan-European standards in the social sciences and chaired the Ethics and Societal Impact Advisory Group for another EC-funded European Demonstration Project on mass transit security (SECUR-ED). He has advised the UK Research Integrity Office; the National Disability Authority (NDA) of the Irish Ministry of Justice; the UK Parliamentary Office of Science and Technology; the Scottish Executive; UK Government Social Research; National Centre for Social Research; the Audit Commission; the Food Standards Agency; the Ministry of Justice; the BIG Lottery; a UK Local Authorities' Consortium; Skills Development Scotland; Agence Nationale de la Recherche (ANR the French Research Funding agency) among many others. Ron was founding Executive Editor of the Emerald gerontology journal *Quality in Ageing and Older Adults*. He published *Ethical Decision Making in Social Research: A Practical Guide* (Palgrave Macmillan, 2009 and 2011) and coedited with Martin Tolich *The SAGE Handbook of Qualitative Research Ethics* (Sage, 2018). He is currently leading a new €2.8M European Commission-funded project (PRO-RES) that aims at promoting ethics and integrity in all non-medical research (2018–2021).

https://roniphofen.com/

ABOUT THE EDITOR

Zvonimir Koporc has international recognition for expertise on research in life sciences (immunology) and professional standards in research ethics. His primary consultative activity in ethics and life sciences at present is for the European Commission (EC) Ethics Unit, Directorate General for Science and Innovation, the Research Executive Agency (REA), and the European Research Council (ERC). He has acted as Consultant, Adviser, and/or delivered training on research ethics at the European and national level. He has worked in several life science teams in Europe, and his PhD in Chemistry was awarded by the Technical University of Vienna, Austria. At the beginning of 2007 he returned to his home country through the national program called "Return of the scientists," which was announced from the Croatian Ministry of Science. As a scientist he took up a position at the prominent Croatian scientific institution – Institute Rudjer Boskovic, Zagreb. Receiving his university tenured track position, he moved then to the Department of Biotechnology, University of Rijeka, where he stayed until 2015. From that year on, he joined Catholic University of Croatia Zagreb where he currently holds a position of University Associate Professor in Physiology where he is also a member of the Ethics Review Board. He has published extensively on immunology. Over and above the life sciences, his current professional interests are in research ethics and scientific integrity. Developing a fruitful cooperation with the Croatian Data Protection Agency in October of 2017, he organized a symposium on "Data protection in research – an insight in to the EU General Data Protection Regulation (GDPR) 25/5/2018" where he acted as a president of the organizing committee. Another fruitful cooperation has been developed with the Croatian National Agency for Mobility and EU Programmes funds where Dr Koporc regularly holds guidance workshops on ethics and research integrity for scientists and research professionals intending to apply for EU funds.

LIST OF CONTRIBUTORS

Joana Araújo is Lecturer and Vice-Director of the Instituto de Bioética, Universidade Católica Portuguesa (Porto, Portugal), where she acts as Research Coordinator and collaborates in several research projects. She holds a doctorate and master's degree in Bioethics awarded from the Instituto de Bioética, Universidade Católica Portuguesa. She is Member of the Office of Ethical Evaluation and Science Integrity at the Portuguese Foundation for Science and Technology and Expert for ethics evaluation of the European Commission. She is also member of the Ethics Committees of the Escola Superior de Saúde de Viseu, Hospital da Luz Arrábida, and Universidade Católica Portuguesa (Porto).

T. Nelson Campbell directed and wrote *PlantPure Nation*. This feature film examines the political and economic factors that suppress information on the benefits of plant-based nutrition, while making the connections of this idea to larger issues such as medical practice, farming, and food deserts. He also leads an organization that he established to organize a grassroots movement around the health message of plant-based nutrition (HealingAmericaTogether.com). In addition, he founded a nonprofit organization (PlantPureCommunities.org) to spearhead a network of hundreds of plant-based support groups, involving (as of the time of this publication) well over 100,000 people, as well as a strategy for bringing nutrition education and affordable foods into underserved communities. Prior to this, Nelson had 25 years of entrepreneurial experience building various companies. He has undergraduate and graduate degrees from Cornell University in political science and economics.

T. Colin Campbell, PhD, has been actively researching diet and health issues for more than six decades. He also spent about 20 of those same years participating in national and international policy development on diet and health. His experimental research initially centered on the demonstration that dietary animal protein sharply increases experimental cancer development in laboratory rodents. Findings from his and other research groups shows that nutrition, provided by whole plant-based foods, creates more health and prevents more disease than all the conventional pills and procedures combined. More generally, it begs the question why nutrition has been so long ignored by the field of medicine.

Ana Sofia Carvalho is Associate Professor with Aggregation in Bioethics, Director of the Instituto de Bioética, and Chair of the Portuguese UNESCO Chair in Bioethics at the Universidade Católica Portuguesa (Porto, Portugal), where she coordinates the Doctoral Program in Bioethics. She holds a doctorate degree in Biotechnology (Escola Superior de Biotecnologia, Universidade

Católica Portuguesa). Prof. Carvalho coordinates the Office of Ethical Evaluation and Science Integrity at the Portuguese Foundation for Science and Technology, and is a Member of the National Ethics Council for the Life Sciences. She is member of the European Group on Ethics in Science and Technology and Expert for ethics evaluation of the European Commission.

Kate Chatfield is Deputy Director of the Centre for Professional Ethics University of Central Lancashire (UCLan), UK. She has a background in philosophy, holistic healthcare provision, and bioethics, and now works as a bioethicist and qualitative researcher across a range of projects. Her main areas of research include global research ethics, animal rights and responsible research and innovation. Previously, Kate was at the forefront of university developments in e-learning, creating innovative postgraduate courses that enabled communal learning for groups of students from all over the world. This included the development of cross-disciplinary modules in research methods, research ethics and critical thinking.

Rok Čivljak is a Specialist in Infectious Diseases and the Head of the Department for Respiratory Tract Infections of the University Hospital for Infectious Diseases in Zagreb, Croatia, also holding the position of the Deputy Hospital Director. He is the Assistant Professor at the University of Zagreb School of Medicine. His major research interests and expertise include acute respiratory infections, healthcare-associated infections, hospital infection control, HIV/AIDS, and infections in travelers and migrants.

Julie Cook is a Research Fellow in the Faculty of Health & Wellbeing at the University of Central Lancashire (UCLan), UK, where she has worked with the Centre for Professional Ethics since 2005. In addition to TRUST, she has contributed to a range of European Commission-funded projects around international justice and ethics in health, science, and technology, specializing and publishing in gender and consent issues. Since 2014, Julie has been co-manager of the Faculty's Academic Research Support Team and is Good Clinical Practice certified. She is Deputy Vice Chair of UCLan's STEMH Ethics Committee.

Kathinka Evers leads philosophy research for the European Flagship Human Brain Project. She is also Senior Researcher and Professor of Philosophy at the Centre for Research Ethics & Bioethics (CRB) at Uppsala University and Professor ad honoram at the Universidad Central de Chile. She has been Invited Professor on the Chair Condorcet at École Normale Supérieure, Paris (2002); at Collège de France, Paris (2006–7); and at Centro de Investigaciones Filosoficas, Buenos Aires (2012). Focusing on philosophy of mind, neurophilosophy, bioethics and neuroethics, Kathinka directs the teaching and research on neuroethics at Uppsala University, where she started the first courses in the subject.

Cristina Gavrilovici, MD, is Associate Professor in Biomedical Ethics at the University of Medicine and Pharmacy, Grigore T. Popa, Iasi, Romania. She graduated the master studies in Bioethics from Case Western Reserve University, Cleveland, Ohio. She holds a PhD degree and a Doctor Habilitatus title. She is member of the Romanian National Ethics Council and serves in several research ethics committees. She has worked as an expert in research ethics for the Ethics Sector of the European Commission since 2006. She is the author of 7 books and 27 books chapters in the field of bioethics. She published over 60 articles in both the bioethics and medicine domain.

James Giordano, PhD, MPhil, is Professor in the Departments of Neurology and Biochemistry and Chief of the Neuroethics Studies Program of the Pellegrino Center for Clinical Bioethics at the Georgetown University Medical Center, Washington DC. He is a Research Fellow of the European Union Human Brain Project; is Chair of the Neuroethics Subcommittee of the *IEEE Brain Initiative*, serves as a consultant in brain science and neurotechnology to the Organisation for Economic Cooperation and Development (OECD); and was an appointed member of the US Department of Health and Human Services Secretary's Advisory Council for Human Research Protection. He has been elected to the European Academy of Science and Arts, and the Dana Alliance of Brain Initiatives; and has been named a Fellow of the Royal Society of Medicine (UK); and a Distinguished Lecturer of the *Institute for Electrical and Electronics Engineers* (IEEE).

Pablo Hernández-Marrero is Senior Researcher and Invited Lecturer at the Instituto de Bioética and CEGE: Centro de Estudos em Gestão e Economia, Universidade Católica Portuguesa (Porto, Portugal). He obtained his doctorate degree in Health Services Organization and Management and his master's degree in Health Sciences: Health Administration from the Institute of Health Policy, Management and Evaluation, University of Toronto (Canada). Dr Marrero has been dedicating the last years of his scientific path to research in palliative care, ethics, and dementia. He also worked in healthcare management, international consultancy, and university teaching. He integrates the Steering Group of the Taskforce on Preparation for practice in palliative care nursing of the European Association for Palliative Care.

François Hirsch graduated in Immunology in the Institut Pasteur and holds a certificate in Science and Medical Ethics from Paris-Sud University. He spent 30 years at the Institut National de la Santé et de la Recherche Médicale (Inserm) occupying various positions, and 3 years at the Unit "Governance and Ethics" in the European Commission (EC) where he helped in setting-up the ethics review of EC-funded proposals. François Hirsch is now member of the Inserm ethics committee and member of several European and national ethics boards.

Monique Ischi has an academic background in international policy and development. Over the last 20 years she has worked professionally in various international environments ranging from field work to policy related activities. Over the last 15 years she has increasingly focused on comprehensive security related issues in her work. In recent years Monique has co-authored articles in the area of dual use, ethics, and biosecurity. She is currently working as an Independent Consultant.

Jasenka Gajdoš Kljusurić is Full Professor at the Faculty of Food Technology and Biotechnology, University of Zagreb, teaching classes in nutrition on the bachelor, diploma and doctoral studies courses. She focuses on ethics and the majority of her scientific papers are in the fields of food control and safety.

Mihalis Kritikos is a Policy Analyst at the European Parliament working as a legal/ethics advisor on Science and Technology issues (STOA/EPRS) and Fellow of the Law Science Technology & Society Programme of the University of Brussels (VUB-LSTS). Mihalis is a legal expert in the fields of EU decision-making, food/environmental law, the responsible governance of science and innovation, and the regulatory control of new and emerging risks. He has worked as a Research Programme Manager for the Ethics Review Service of the European Commission, as a Senior Associate in the EU Regulatory and Environment Affairs Department of White and Case (Brussels office), as a Lecturer at several UK Universities, and as a Lecturer/Project Leader at the European Institute of Public Administration (EIPA). He also taught EU Law for several years at the London School of Economics and Political Science (LSE). He holds a Bachelor in Law (Athens Law School), master's degrees in European and International Environmental Law and Environmental Management (University of Athens and EAEME respectively), and a PhD in European Risk Regulation (London School of Economics-LSE). In 2008, he won the UACES Prize for the Best Thesis in European Studies in Europe.

Albena Kuyumdzhieva works in the Ethics and Research Integrity Sector, Scientific Advice Mechanism Unit at DG Research and Innovation, European Commission. Albena is a certified data protection officer whose work is focused on ensuring ethics compliance of personal data processing as part of the Horizon 2020 Ethics Review Process. In this capacity, Albena has been involved in drafting policy guidelines and recommendations related to ethics and data protection in research context. Before joining the European Commission, Albena has provided advisory services to 16 governments in the Eastern European Neighbourhood Policy Region, the Balkans, Central Asia and West Africa. Albena holds the degree of Doctor of Philosophy and master's degrees in Law and Finances.

Janet Mifsud is a Member of the Department of Clinical Pharmacology and Therapeutics, University of Malta. Her area of expertise is in the pharmacology of drugs used in epilepsy, toxicology, and also ethical issues in health. She is a Fulbright Scholar and is involved in several EU-wide research projects. She was formerly Vice-President (Europe) International Bureau for Epilepsy. She has a degree in Theology, has sat on several national ethics committees, and has been chosen several times by the European Commission to contribute as external expert evaluator for Ethics Panels. She sits on several National Committees such as the President's Foundation for Social Well Being and also as a board member of the Malta Council for Science and Technology. She has published extensively in her area of expertise.

Lluis Montoliu (Barcelona, 1963) is a Biologist working at the National Centre for Biotechnology (CNB-CSIC) in Madrid since 1997, member of the Spanish initiative on Rare Diseases (CIBERER-ISCIII), and Director of the Spanish node of the European Mouse Mutant Archive (EMMA/INFRAFRONTIER). His laboratory has generated numerous animal models of human rare diseases, such as albinism, through standard genetic modifications or using CRISPR-Cas9 genome-editing tools. He is a member of the CSIC Ethics Committee and of the ERC Ethics Panel. He founded the International Society for Transgenic Technologies (ISTT) and is now the President of the European Society for Pigment Cell Research (ESPCR). He is currently leading the Association for Responsible Research and Innovation in Genome Editing (ARRIGE) initiative.

Sandra Martins Pereira is Senior Researcher and Invited Lecturer at the Instituto de Bioética and CEGE, Universidade Católica Portuguesa (Porto, Portugal), where she is Principal Investigator of Project InPalIn: Integrating Palliative Care in Intensive Care. In 2013, she was awarded with a Marie Curie International Training Network post-doctoral research fellowship in palliative and end-of-life care research at the Vrije Universiteit Medisch Centrum and EMGO+ Institute for Health and Care Research in Amsterdam (Netherlands). She holds a doctorate and master degree in bioethics awarded from the Instituto de Bioética, Universidade Católica Portuguesa. She is screening editor and member of the Editorial Board of *Palliative Medicine* (SAGE journals).

Nicola Petrosillo has a degree in Medicine (1977), and specialization in Infectious Diseases (1981) and Internal Medicine (1985). Currently, he is Director of the Clinical and Research Infectious Disease Department at the National Institute for Infectious Diseases "Lazzaro Spallanzani" in Rome. He is President of Infection Control Multidisciplinary Joint Committee of UEMS. He is co-leader of the ESCMID Emerging Infections Taskforce. He was WHO clinical and IPC consultant in Lagos, Nigeria, for the 2014 Ebola epidemic, running the hospital for Ebola patients. He has been President of the Italian Society for Healthcare Associated Infections (SIMPIOS). He is Editor-in-Chief of *Infectious Disease Reports*.

Johannes Rath has academic background in molecular biology, toxicology, and law. He has worked for 20 years as staff scientist at the University of Vienna and published extensively on dual use, security, and safety-related issues. He also worked as Chief Inspector for the United Nations Special Commission in Iraq and is currently responsible for all safety-related issues of an international organization. Over the last 15 years he has worked as independent ethics and science advisor to various organizations and multinational projects.

Doris Schroeder is Director of the Centre for Professional Ethics, University of Central Lancashire (UCLan), UK, and Professor of Moral Philosophy in the

School of Law at UCLan Cyprus. Her academic background is in philosophy, politics, and management/economics. She has coordinated several international EU projects, including TRUST, and acts regularly as a consultant for the European Commission. Doris' main areas of research interest are in human rights, responsible research and innovation, global research ethics and conceptual work, for instance, on dignity, justice, and vulnerability. She is the lead author of the *Global Code of Conduct for Research in Resource-Poor Settings*, developed by TRUST.

INTRODUCTION: RESEARCH PRODUCTION IN LIFE SCIENCES

Zvonimir Koporc

THE RESEARCH PRODUCTION PROCESS IN LIFE SCIENCES

I would like to open this fourth volume in the *Advances in Research Ethics and Integrity Series* with some personal impressions about life sciences research based on my experience working in this field. Life science research teams in many ways resemble small factories. In factories, each part of the production chain is dedicated to another and each part of the production chain is strongly dependent on the efficiency of a preceding one. Just as with industrial factories, life sciences research teams are pressed with upcoming deadlines, productivity output (measured in terms of both patents and publications), and a constant struggle to finance their activities. Life science research is costly and often strongly dependent on very expensive chemicals and high-tech equipment. In such a competitive and challenging environment, only the most efficient research teams can survive.

Some might not see that as an issue of ethics or of integrity, but it is, and it is more than just a problem of the additional stress and pressure on the team members. Life science more often than not has to deliver a solid product in terms of research results, a patent, a treatment, a vaccine, a therapeutic model, or similar, and the team has to think and act together as a small quasi-commercial, and often commercial, enterprise. This can make for a more rigid decision-making process – the need to hone and refine the research "products" and ensure that obligations to funders are met. In other research fields, such as the social sciences or humanities, there may appear to be greater freedom for more "independent" thinking about research directions – although there may be similar constraints to "deliver" the outcomes specified by contractors – none more so than in market research.

Ethics and Integrity in Health and Life Sciences Research
Advances in Research Ethics and Integrity, Volume 4, 1–9

ISSN: 2398-6018/doi:10.1108/S2398-601820180000004001

As in the preceding volumes in this Series, this all points to how vital the institutional research infrastructure is to the integrity of the scientists involved. The ability to pursue particular lines of thought is intricately linked to organizational pressures and the individual career trajectory of the researcher.

Particularly, costly elements in life sciences involve the fundamental resources needed to even begin the research. To give an example, the annual price of monoclonal antibody (mAb) therapies is about $100,000 higher for research in oncology and hematology than for researching other disease states (Hernandez et al., 2018), making research in these fields even more expensive. Different research kits such as ELISA, biochemical assay kits, or Flow cytometry kits, together with the equipment purchase and its maintenance, can sometimes exhaust the whole planned annual research budget.

Such costs do threaten research integrity since the "temptations" to cut corners or ensure positive results when they are not clear are high. Certainly, governmental financial support allocated for life sciences is strong in most of the well-developed European Union (EU) countries. Without that, it would be almost impossible to build an efficient life science research infrastructure. Nevertheless, those allocations are often not sufficient to cover all necessary research costs. The recent economic crisis aggravated that already difficult situation slightly declining the percentage of the share of research and innovation (R&I) in total public budgets for the EU as a whole (1.4% 2014 vs. 1.5% 2007).[1] There is a growing tendency to expose life science research to the real market conditions globally decreasing governmental support for research and development (R&D).[2]

That is why even in public research institutions, life scientists are frequently forced to search for additional financial resources to support their projects.

RESEARCH WITH ANIMALS

Having worked a lot with animal research models, my own perceptions have been particularly influenced by the demands of such research. Research using animals is usually one of the most expensive forms of scientific research. Above the purchase of such animal models, the cost of their housing and the maintenance of animal facilities are factors which require careful planning and experienced project management. Besides the expense and the technical demands which follow such research, a skilled staff with a strong ability to quickly recognize and effectively respond to all ethical issues which can emanate from their research is essential for the appropriate implementation of proposed work.

Animal welfare is very important for most EU citizens as shown in the results of a recent Eurobarometer study,[3] but citizens' feelings of not being well informed about that topic can often give rise to a negative perception of all research performed with and on animals.[4] As mentioned in the introduction to this Series, different rights, values, and principles often stand in opposition to each other. Thus, recent global actions for the fight against terrorism and the seeking of security can stand in opposition to rights of privacy (Iphofen, 2014). In similar vein, the growing concerns for animal rights may oppose to some degree the right to do research and gain knowledge in a common public interest such as a cure for specific diseases. From the time when Russell and Burch's *The Principles of Humane Experimental Technique* was first published in 1959, in which they proposed a new applied science that would improve the treatment of laboratory animals (the 3Rs principle – **R**eplacement, **R**eduction, and **R**efinement), to nowadays when the interpretation of that principle has gone through several "updates" (Tannenbaum & Bennett, 2015), pros and cons for animal testing still stand as a hot topic for the public and for scientists alike.

For research teams working on laboratory animals, a constant struggle with often unfair and oversimplified public perceptions which sometimes places them into the category of animal rights "offenders" is another challenging factor in their endeavors. Such oversimplified views of researchers' hard work and engagements are frustrating when researchers have to go through rigorous training to obtain certificates that permit them to work with laboratory animals. Those certificates are designed to prove their knowledge and technical skills, but also their awareness of the ethical issues associated with research performed on laboratory animals.

Recently I spoke with the director of one prominent research center in the EU about that public perception. His comment demonstrates how wrong he believes that perception to be:

> Even if someone would not believe our training certificates, ethical approvals, our empathy for those animals and our noble intentions, I am sure that the public will understand that we treat our animals in the best way and take the maximal care of them also because those animals are our precious and very expensive research tools.

Animal researchers are constantly working on the improvement of already established ethical practices for research on animals.[5] For instance, a report from the Canadian Animal Care Committee demonstrated the need for the scientific community to make progress concerning the implementation of "refinement" and "reduction", particularly in the creation and use of genetically engineered animals; addressing these roadblocks needs to be a priority (Ormandy, Dale, & Griffin, 2013). Another publication showed that a web-based Nano technical summary database called "AnimalTestInfo" offers the possibility for scientists to do large-scale analyses of planned animal studies to inform researchers and the public (Bert et al., 2017). Data drawn from that database can provide a guidance for rethinking the role of animal research models.

In regard to the large animal research models, a recent Review of Research Using Non-Human Primates (NHP) founded by the UK Biotechnology and Biological Sciences Research Council (BBSRC), Medical Research Council (MRC), and Wellcome Trust from the Panel involving prominent experts in different life science fields recommended that in their public engagement, the funders and researchers should avoid overstating and generalizing the medical benefit of NHP research, since this cannot be substantiated in many cases.[6] It is obvious that the research community is constantly investigating the need for use of the animals in different research areas. Yet, some animal researchers don't even tell their own families about their work for fear of attack by extremists.[7]

ETHICS CODES FOR THE LIFE SCIENCES

Although a code of ethics for the life sciences (Jones, 2007) and ethical review practices concerning research with animals are well established, a dynamic and a balanced approach is needed. The approach must be "dynamic" to be able to modify and adopt the best practices and "balanced" to focus on the important and essential parts of each sets of rights. Only by implementing those two principles can the potential contrast between those rights be reconciled.

The workload and complexity of research in life science research teams requires some way of standardization which can allow the "symbiosis" between high productivity and process simplification. In such teams, each research process step and each individual research role is clearly defined. That work organization could enable high research productivity; some tasks and rules are prepared generally for the whole team. Whether this is the question of the work safety, patent and data protection, or in-vivo research on laboratory animals, the idea of such "generalized rules" is to facilitate, accelerate, and increase the efficiency of the whole research process. It is expected that each individual member will follow such rules. This means that to be productive, life science teams must be highly arranged and coordinated systems. Nevertheless, errors can happen anywhere, and life science research teams are not immune.

In such set-ups, there are two ways rule violations can happen both on the team and on the individual level. Individual violations are to be expected due to human characteristics – the flaws of self-interest, ambition, greed, and so on. The consequences of such violations are more predictable and as such could potentially be recovered if recognized in time and if the organizational controls are effective. More problematic violations are those that arise out of the organizational infrastructure which affect the whole team – whether deliberate or latent. In that case consequences are harder to address as they lead to the establishment of unwanted or unlawful practices. These influences can easily go unrecognized in some countries in which ethics review and integrity practices are not efficiently set or where they are in their early stages.

ETHICS DUMPING IN LIFE SCIENCES

The export of unethical practices from high-income regions to low- and middle-income countries is one of the critical ethical concerns we still face in the twenty-first century. This practice was recently officially termed by the European Commission as "ethics dumping" (EC, 2016). That burning issue is the core topic of the EU project TRUST which is discussed in Chapter 1 in this volume. Although problems arising from research conducted from the wealthier environments in resource-poor settings were recognized in the past (Molyneux et al., 2009), the attention to the ethics dumping is rapidly growing (Schroeder, Cook, Hirsch, Fenet, & Muthuswamy, 2018). Any research carried in a resource-poor setting brings many challenges. For instance, obtaining consent for research in such settings requires additional considerations. That is why the consenting process should be strengthened by taking into account local social, cultural, and economic contexts in the design and administration of consent forms and processes (Boga et al., 2011). The TRUST project aimed to develop a Global Code of Conduct for Research in Resource-Poor Settings which will provide a guidance across all research disciplines. The challenges of resource-poor settings we also discuss in Chapter 2, where we investigate those settings from the viewpoint of providing health care during the largest ever outbreak of Ebola virus disease (EVD), which started during the 2013 in West African countries. This interesting chapter also provides the author's personal impressions of the situation in Lagos, Nigeria, where Dr. Petrosillo was deployed as a World Health Organization (WHO) clinical consultant on EVD. In this chapter, the author provides us with a clear table containing the issues and consequences of ethical considerations associated with dealing with EVD. The chapter clearly concludes that the epidemic would have been far worse and more people would have died without the provision of international help.

DATA PROTECTION AND THE USE OF BIG DATA IN LIFE SCIENCES

As stated previously in the Series Preface with which Iphofen opens this volume, the uneven distribution of ethics standards across whole European Research Area (ERA) (Evers, 2004) has led to the recognition of a need for a harmonization between improved review practices with the existing ones. The intention of this volume is to ensure support to life scientists, provide an overview of ethics questions and issues in some specific life science fields, and inform a broad audience about the current challenges and ongoing projects related to ethics in the life sciences. A rapid development of new technologies, expeditious development of the digital revolution and new media has also entered the field of life sciences and biomedicine. An overview of the ethics challenges in the digital era focusing on medical research in relation to the new EU General Data Protection Regulation (GDPR) 2016/679 is presented in Chapter 3. This new regulation replaces the Data Protection Directive 95/46/EC and is designed to harmonize data privacy laws across Europe, to protect and empower all EU citizens' data privacy, and to reshape the way organizations across the region approach data privacy (FRA/ECtHR/EDPS, 2018).

Huge amounts of information about patients' health, so-called Big Data, carry a tremendous potential for use in life science research even for treating the most severe diseases. It has been recognized that Big Data, as a paradigm, can be a double-edged sword in organizational and management research, capable of significantly advancing our field but also causing backlash if not utilized properly (Wenzel & Van Quaquebeke, 2018). Furthermore, the right for privacy and right for research on common benefit are in opposition. This susceptible relation is explored in Chapter 4.

IN SITU ETHICS, A FOCUS ON LABORATORY WORK

Sometimes the interpretations of principles of safety and security are very limited, and are considered only as sub-criteria of other ethical principles, ignoring the public interest in safety and security. That issue we discuss in Chapter 5 where it is clearly recognized that as long as adequate inclusion of safety and security expertise in ethics panels is missing, as ethics panels are sometimes dominated by philosophers, ethicists, medical doctors, and lawyers with limited practical background in safety and security risk management of emerging technologies, ethics reviews will not contribute to effective safety and security risk management when emerging technologies are involved.

One of such new emerging technologies which will have a strong impact both on science and society in the future is gene-editing technology. The legal and ethical considerations governing gene-editing technology in the European Union are scrutinized in Chapter 6. Recently, a need for the foundation of an expert group (European Steering Committee) to assess the potential benefits and drawbacks of genome editing and to design risk matrices and scenarios for a responsible use of this promising technology was recognized (Chneiweiss et al., 2017). Consequently, on March 23, 2018, in Paris, France, the international "Association for Responsible Research and Innovation in Genome Editing" (ARRIGE) to promote a global governance of genome editing was launched.[8] The association aims to provide a comprehensive setting for different stakeholders, including academia, private sector, patient organizations, public opinion, and governmental institutions, to allow the development of these paramount technologies in a safe and socially acceptable environment. Recognizing the importance of this event, we dedicated a separate chapter (Chapter 7) on the ARRIGE association.

Above the issue of security and safety of new and emerging technologies, the so-called dual-use issue plays a significant ethical consideration for technologies in which the future consequences of their use are not clearly predictable. The concept of dual use can be applied on almost everything that is designed and produced. Nevertheless, this concept in life sciences may have even deeper meaning, since the research products that life sciences may enter have an application already at the nanoparticle level, spreading such applications to a fully global use.

The need for a better definition of the dual-use concept, with the need to move from constraining to enabling types of policies, the move from secrecy to openness, and the move from segregation to integration of the public voice, has been previously recognized (Dubov, 2014). Also a reconsideration of dual use in life sciences should include the aspect of threats and intentions (van der Bruggen, 2012).

In Chapter 8, a dual use with focus on neuroscientific and nanotechnological research is elaborated in relation to the current and emerging tools and techniques used in brain research, presenting how they actually or potentially can be employed in settings that threaten public health and raising multiple ethical concerns. A term like neurohacking (Malin et al., 2017) is exhaustively elaborated in several research domains.

A problem of the rapidly increasing numbers of older adults with dementia and other mental health problems throughout the world will have huge implications for healthcare systems (Sorrell & Loge, 2010), consequently making clinical dementia research more demanding and more urgent.

In Chapter 9 we moved a step ahead, exploring the ethical challenges of informed consent, decision-making capacity, and vulnerability in clinical dementia research where we tried to provide an overview of the ethical framework and decision-making in clinical dementia research, and to analyze and discuss the ethical challenges and issues that can arise when conducting clinical dementia research.

Above multiple ethical challenges which are raised in this kind of research, even the nutrition of such patients must be in focus for ethical considerations. For instance, a comfort feeding (not forcing a person to eat or making them feel guilty if they don't), instead of enteral nutrition, can and must be applied when dysphagia and progressive disease processes have still not occurred. Enteral nutrition guidelines and recommendations which have been developed by the American Society for Parenteral and Enteral Nutrition and the Academy of Nutrition and Dietetics for individuals with dementia focus on patients with advanced dementia due to the dysphagia and progressive disease process. Nevertheless, the application of the enteral nutrition to all dementia patients is still widely practiced (Schwartz, 2018).

Malnutrition problems in the broad population of older adults is a growing issue, and a comprehensive assessment for early identification of malnutrition and determination of the appropriate intervention strategies is needed to address this global public health problem (Maseda et al., 2018).

According to the Second International Conference on Nutrition (ICN2), which was jointly organized by the Food and Agriculture Organization of the United Nations (FAO) and the World Health Organization (WHO) and was held at the FAO Headquarters in Rome, Italy, from 19 to 21 November, 2014, 795 million people remain undernourished, over 2 billion people suffer from various micronutrient deficiencies, and an estimated 161 million children under 5 years of age are stunted, 99 million underweight, and 51 million wasted. Meanwhile, more than 600 million adults are obese. Global problems require global solutions (Amoroso, 2016).

Recognizing raising global health problem related to nutrition, we included in this volume two additional chapters: one describing the ethical and moral responsibility related to diet therapy (Chapter 10) and another, bolder chapter which tries to move a step further, moving from the "conventional box" of relations between nutritional and medical practices (Chapter 11).

In this volume, we have tried to identify some of the most important crossroads between the life sciences and the issues of ethics and integrity they must confront. Certainly, this task was not easy given the broad range of research fields which appertain to life sciences. However, we did not want to renege upon that challenging adventure. As each great journey starts with a first step, we hope that this volume will encourage professionals in this field in their professional endeavors to further their work with integrity. We also hope this volume offers some insight on the ethical problems facing the life sciences for the interested general reader. Each chapter here warrants another volume of its own and will be returned to later in the series.

NOTES

1. Retrieved from https://ec.europa.eu/research/openvision/pdf/rise/veugelers-saving_away.pdf
2. Retrieved from https://digital.rdmag.com/researchanddevelopment/2018_global_r_d_funding_forecast
3. Retrieved from http://ec.europa.eu/public_opinion/archives/ebs/ebs_270_en.pdf
4. Ipsos MORI Social Research Institute. Public attitudes to animal research in 2016. A report for the Department for Business Innovation & Skills. July 2016. Retrieved from https://www.ipsos.com/sites/default/files/publication/1970-01/sri-public-attitudes-to-animal-research-2016.pdf
5. http://digital.rdmag.com/researchanddevelopment/2018_global_r_d_funding_forecast?pg=1#pg1
6. https://wellcome.ac.uk/sites/default/files/wtvm052279_1.pdf
7. Retrieved from https://www.theguardian.com/education/2007/feb/13/businessofresearch.highereducation
8. Retrieved from https://arrige.org/

REFERENCES

Amoroso, L. (2016). The Second International Conference on Nutrition: Implications for Hidden Hunger. Hidden Hunger: Malnutrition and the First 1,000 Days of Life: Causes, Consequences and Solutions. H. K. Biesalski and R. E. Black. Basel, Karger. *115*, 142–152. Retrieved from https://www.ncbi.nlm.nih.gov/pubmed/27197665

Bert, B., Dorendahl, A., Leich, N., Vietze, J., Steinfath, M., Chmielewska, J., … Schonfelder, G. (2017). Rethinking 3R strategies: Digging deeper into animal test info promotes transparency in in vivo biomedical research. *PLoS Biology*, *15*(12), e2003217.

Boga, M., Davies, A., Kamuya, D., Kinyanjui, S. M., Kivaya, E., Kombe, F., … Mwalukore, S. (2011). Strengthening the informed consent process in international health research through community engagement: The KEMRI-Wellcome Trust Research Programme Experience. *PLoS Medicine*, *8*(9), e1001089.

Chneiweiss, H., Hirsch, F., Montoliu, L., Muller, A. M., Fenet, S., Abecassis, M., … Saint-Raymond, A. (2017). Fostering responsible research with genome editing technologies: a European perspective. *Transgenic Research*, *26*(5), 709–713.

Dubov, A. (2014). The concept of governance in dual-use research. *Medicine, Health Care and Philosophy*, *17*(3), 447–457.

Evers, K. (2004). Codes of conduct – Standards for ethics in research. European Commission Luxembourg: Office for Official Publications of the European Communities, 2004. Retrieved from https://www.researchgate.net/publication/301609039_Codes_of_Conduct_Standards_for_ethics_in_research

FRA/ECtHR/EDPS. (2018). Handbook on European data protection law. Retrieved from https://www.echr.coe.int/Document/Handbook_data_protection_02ENG.pdf

Hernandez, I., Bott, S. W., Patel, A. S., Wolf, C. G., Hospodar, A. R., Sampathkumar, S., & Shrank, W. H. (2018). Pricing of monoclonal antibody therapies: higher if used for cancer? *American Journal of Managed Care*, *24*(2), 109–112.

Iphofen, R. (2014). Ethical issues in surveillance and privacy, Chapter 5. In A. W. Stedmon & G. Lawson (Eds.), *Hostile Intent and Counter-Terrorism: Human Factors Theory and Application* (pp. 59–71). Aldershot: Ashgate.

Jones, N. L. (2007). A code of ethics for the life sciences. *Science and Engineering Ethics*, *13*(1), 25–43.

Malin, C. H., Gudaitis, T., Holt, T. J., Kilger, M., Malin, C. H., Gudaitis, T., Holt, T. J., ... Kilger, M. (2017). *Looking Forward: Deception in the Future*. London: Academic Press Ltd-Elsevier Science Ltd.

Maseda, A., Diego-Diez, C., Lorenzo-Lopez, L., Lopez-Lopez, R., Regueiro-Folgueira, L., & Millan-Calenti, J. C. (2018). Quality of life, functional impairment and social factors as determinants of nutritional status in older adults: The VERISAUDE study. *Clinical Nutrition*, *37*(3), 993–999.

Molyneux, C., Goudge, J., Russell, S., Chuma, J., Gumede, T., & Gilson, L.. (2009). Conducting health-related social science research in low income settings: ethical dilemmas faced in Kenya and South Africa. *Journal of International Development*, *21*(2), 309–326.

Ormandy, E. H., Dale, J., & Griffin, G. (2013). The use of genetically-engineered animals in science: perspectives of Canadian Animal Care Committee members. *Alternatives to Laboratory Animals*, *41*(2), 173–180.

Schroeder, D., Cook, J., Hirsch, F., Fenet, S., & Muthuswamy, V.. (2018). *Ethics dumping case studies from North-South research collaborations*. Springer Briefs in Research and Innovation Governance. Retrieved from https://www.springer.com/gb/book/9783319647302

Schwartz, D. B. (2018). Enteral nutrition and dementia integrating ethics. *Nutrition in Clinical Practice*, *33*(3), 377–387.

Sorrell, J. M., & Loge, J. (2010). Implications of an aging population for mental health nurses. *Journal of Psychosocial Nursing and Mental Health Services*, *48*(9), 15–18.

Tannenbaum, J., & Bennett, B. T. (2015). Russell and Burch's 3Rs then and now: The need for clarity in definition and purpose. *American Association for Laboratory Animal Science*, *54*(2), 120–132.

van der Bruggen, K. (2012). Possibilities, intentions and threats: Dual use in the life sciences reconsidered. *Science and Engineering Ethics*, *18*(4), 741–756.

Wenzel, R., & Van Quaquebeke, N. (2018). The double-edged sword of Big Data in organizational and management research: A Review of Opportunities and Risks. *Organizational Research Methods*, *21*(3), 548–591.

PROMOTING EQUITY AND PREVENTING EXPLOITATION IN INTERNATIONAL RESEARCH: THE AIMS, WORK, AND OUTPUT OF THE TRUST PROJECT

Julie Cook, Kate Chatfield and Doris Schroeder

ABSTRACT

Achieving equity in international research is one of the pressing concerns of the twenty-first century. In this era of progressive globalization, there are many opportunities for the deliberate or accidental export of unethical research practices from high-income regions to low- and middle-income countries and emerging economies. The export of unethical practices, termed "ethics dumping," may occur through all forms of research and can affect individuals, communities, countries, animals, and the environment. Ethics dumping may be the result of purposeful exploitation but often arises from lack of awareness of good ethical and governance practice.

This chapter describes the work of the TRUST project toward counteracting the practice of ethics dumping through the development of tools for the improvement of research governance structures. Multi-stakeholder consultation informs all of TRUST's developments. Most importantly, this gives voice to marginalized vulnerable groups and indigenous people, who have been equal and active partners throughout the project.

At the heart of the TRUST project is an ambitious aim to develop a Global Code of Conduct for Research in Resource-Poor Settings. *Uniquely, the Code provides guidance across all research disciplines in clear, short statements,*

Ethics and Integrity in Health and Life Sciences Research
Advances in Research Ethics and Integrity, Volume 4, 11–31

ISSN: 2398-6018/doi:10.1108/S2398-601820180000004002

focusing on research collaborations that entail considerable imbalances of power, resources and knowledge and using a new framework based on the values of fairness, respect, care, and honesty. The code was recently adopted by the European Commission as a reference document for Horizon 2020 and Horizon Europe.

Keywords: International research; ethics dumping; ethics governance; international justice; global ethics; marginalized vulnerable group

INTRODUCTION

The progressive globalization of research activities has resulted in an ever-increasing number of transnational studies (Gainotti et al., 2016; Ravinetto et al., 2016). Joint ventures between multiple stakeholders from different countries are commonplace in all forms of research, and a growing number of researchers from high-income countries (HICs) are electing to conduct their research activities in low- and middle-income countries (LMICs) (Glickman et al., 2009). A number of potential incentives and motivations for such collaborations have been proposed. For the LMIC partners, these include access to funding and other resources that might not be available otherwise (Bradley, 2017). For the HIC partners, cooperative ventures may convey operational and/or economic advantages (Dickson, 2006; Luna, 2009).

While such collaborations may yield benefits for both partners, they may also provoke sensitive ethical issues. Ethical review processes, compliance structures, and follow-up mechanisms can differ greatly between partner countries. Consequently, there is a risk that research that is not permissible in a HIC will be exported to those LMICs where the legal and regulatory frameworks for research are not as rigorous. The European Commission (EC) has recently termed this practice "ethics dumping" (European Commission, 2016).[1] The challenges for cross-cultural research, undertaken in resource-poor settings by researchers from wealthier environments, have long been recognized (Molyneux et al., 2009), but the practice of ethics dumping is receiving a growing amount of attention (Novoa-Heckel, Bernabe, & Linares, 2017; Schroeder, Cook, Hirsch, Fenet, & Muthaswamy, 2018). The European Union (EU) is currently funding actions to address the risk of ethics dumping from both public and private research (European Commission, 2016). One such action is the EU-funded project TRUST.[2]

The goal of the TRUST project is to catalyze a global effort to improve adherence to high ethical standards in research around the world. In an interdisciplinary, global collaboration with 13 partners including multi-level ethics bodies, policy advisors, civil society organizations, funding organizations, industry, academic scholars from a range of disciplines, and representatives from vulnerable research populations, TRUST combines long-standing, highly respected efforts to build international research governance structures. The project's main strategic output consists of three tools to help counteract the practice of ethics dumping:

(1) a fair research contracting online tool;
(2) a compliance and ethics follow-up self-appraisal tool; and
(3) a global code of conduct for research in resource-poor settings.

The fair research contracting tool is an interactive online tool that is designed to assist LMIC partners in making contractual demands on HIC partners without the need for their own specialist legal teams. It focuses on issues such as the fair distribution of post-research benefits, intellectual property rights, data, and data ownership. The compliance and ethics follow-up self-appraisal tool is the component of TRUST's recommended approaches to ensuring compliance with research ethics requirements in LMICs beyond the ethics approval stage. The tool is intended to be practical, accessible, at little or no cost to the intended users.

Both of these tools are vital components of TRUST's activities to counteract ethics dumping, but in this chapter, we spotlight the development of the third tool, namely, a *Global Code of Conduct for Research in Resource-Poor Settings*. It is anticipated that researcher adherence to this innovative code will reduce the prospect of ethics dumping significantly. Crucially, funders can promote adherence to the Code by adopting it as a requirement for funding of collaborative research that is undertaken in resource-poor settings.

Given its injurious and pervasive nature, the practice of ethics dumping is the central motivator for TRUST and, in keeping with the steps taken in our development of the Code, we begin with consideration of the nature and extent of ethics dumping, including real-world examples from LMICs.

ETHICS DUMPING

> To be vulnerable means to face a significant probability of incurring an identifiable harm, while substantially lacking the ability or means to protect oneself. (Schroeder & Gefenas, 2009, p. 117)

The term "dumping" has been traditionally used to describe predatory pricing policies (Investopedia, 2018). In this sense, it refers to the export by a country or company of a product at a price that is lower in the importing market than the price charged in the domestic market; the practice is intentional, with the primary purpose of obtaining a competitive advantage in the foreign market. In the context of research ethics, it has similar connotations, and one can speak of ethics dumping in mainly two areas. First, when research participants and/or resources in LMICs are exploited *intentionally*, for instance because research can be undertaken in an LMIC that would be prohibited in a HIC. Second, exploitation can occur due to insufficient ethics awareness on the part of the researcher, ethics committees in their institutions, or low research governance capacity in the host nation.

For instance, a European researcher might accept a thumbprint on an informed consent document from an illiterate, indigenous research participant in a resource-poor setting and assume this is adequate. However, there may be an ethics infrastructure in the country from which ethics approval should be sought, and additionally the wider community may have already set up a protocol for community assent of research projects prior to any individual informed consent being sought. Even if it has not, it is likely to have its own customs or preferences for authorizing such activities. Observance of such may not normally be required in European settings, or by European ethics approval systems, but it is often both practically and technically essential, as well as ethical, to obtain input from community leaders before enrolling highly vulnerable people in research studies.

Both extreme[3] and moderate[4] poverty increase the likelihood that communities and individuals will be exploited. The international debate on bioethics has long noted the existence of "double standards" (Macklin, 2004), and observed that advantage is being taken of vulnerable people in vulnerable nations. However, while there is global debate about exporting unethical business practices (e.g., bribery and corruption, or tax avoidance) and unethical clinical trials, there is currently little global debate about research in general, or providing guidance to researchers from different disciplines.

The ethics of multidisciplinary research is complex as there are variances in vocabulary and the language of medical ethics is often transferred to other areas unsuccessfully. For example, the ethical implications of "incidental findings" in research are easily understood in the biomedical field. When unexpected, health-related information comes to light over the course of a study, researchers are confronted with ethical dilemmas about whether participants should be informed. However, in other fields, "incidental findings" may have a completely different meaning, with distinctive implications and demand a different kind of ethical analysis.

Global and multidisciplinary collaborations can lead to confusion for all stakeholders about which governance structures and legal instruments are applicable to them. While research involving clinical trials has received considerable attention for several decades, it is often unclear how non-medical research should be governed, especially in borderline areas (e.g., food research involving human participants) or across cultural differences (e.g., different views on animal welfare). Yet:

> [...] for vulnerable populations in developing countries it makes no difference whether they are exploited by an anthropologist or a genetic researcher. They all take something, they shouldn't have taken and then leave, whether it is knowledge, opinions or biological samples is irrelevant. (Anonymous consortium member from the South African San Institute)

ETHICS DUMPING CASE STUDIES

> One can speak of exploitation when we treat ...[others'] vulnerabilities as opportunities to advance our own interests or projects. It is degrading to have your weaknesses taken advantage of, and dishonourable to use the weaknesses of others for your ends. (European Commission, 2010, p. 127)

Multi-stakeholder engagement informs all of TRUST's developments, and one of the first project activities was to identify and analyze real cases of ethics dumping. To this end, a fact-finding workshop was held in Mumbai in March 2016 with a range of participants with responsibility for ethics governance in health research across India. India has experienced many cases of ethics dumping, but also has a sophisticated and developing ethics oversight and governance system, so is ideally placed to identify input for TRUST. At this meeting, participants discussed more than 50 cases from India demonstrating the breadth of ethics dumping concerns, including:

- a project to demonstrate the HPV (Human papilloma virus) vaccine in teenage girls with concerns about legitimate consent and serious adverse event (SAE) reporting,
- participants volunteering for clinical trials in order to obtain health care,
- lack of post-trial access to successful treatments for participants in clinical trials,
- NGO research undertaken without ethical governance due to lack of committee jurisdictions,
- genomics research carried out by teams from overseas among tribal populations without Indian ethical review which raised concerns about reuse of samples, commercial exploitation, and benefit sharing, and
- problems with equity and authorship for Indian researchers in research publications.

In addition to this workshop, TRUST launched an international case study competition in December 2015 to collect global case studies of both ethics dumping in collaborative research and good practice to counter it. Some of the submissions have been included in the anthology *Ethics Dumping* (Schroeder et al., 2018), and a selection of these cases, indicating key areas for concern, are summarized below.

Cervical Cancer Screening in India

Three clinical trials took place in India between 1998 and 2015 in urban and rural areas of India: Mumbai, Osmanabad, and Dindigul. All the women recruited were poor and socially disadvantaged, without universal access to health care in areas where cervical cancer was known to be of high incidence and prevalence. The trials aimed to determine whether trained healthcare workers could conduct cervical cancer screening in the community using cheap methods of testing – primarily visual inspection of the cervix with acetic acid (VIA) – to reduce the incidence and mortality rate of cervical cancer.

The clinical trials were conducted on approximately 347,000 women, of whom about 141,000 were placed in the control arm (no screening). They were provided with so-called usual care or standard care, consisting of health education on cervical cancer symptoms, screening, and treatment, and the availability of local facilities. The standard of care for testing of the disease in India has been cytology screening (Pap smear as per the international standard) since the

1970s, but screening for cervical cancer is not available universally under a government program, although it is available in all major hospitals. The standard of care was therefore misconstrued to be no screening at all.

The women in the trial were observed (mainly retrospectively through medical records) to determine how many would get cervical cancer and how many would die, if they were never screened. This placed them at a known risk of developing invasive cervical cancer, and dying from it, because it was not detected and treated in time. Of women in the no-screening arm of the trial, 254 died due to cervical cancer as per the latest published reports.

As these trials were non-drug-related, prior permission from the Indian government in accordance with guidelines was not required at the time. The regulatory authorities involved were institution-based and unaccountable to legal oversight. A no-screening arm would not have been permitted in the USA, or in France, but was accepted for these clinical trials in India by the US sponsor (National Institutes of Health [NIH]), and the collaborator in France (the International Agency for Research on Cancer [IARC]), a specialized agency of the World Health Organization (WHO). US regulatory authorities claimed an inability to act on complaints regarding the Osmanabad and Dindigul trials as these were funded by a private foundation (Bill & Melinda Gates Foundation [BMGF]) and applied a retrospective waiver of the need for informed consent for the Mumbai trial.[5]

An International Collaborative Genetic Research Project Conducted in China

Between 1994 and 1998 a research team from a renowned US university in receipt of substantial NIH and biopharmacy company funding collected blood and DNA samples from tens of thousands of farmers in the economically disadvantaged province of Anhiu in China, under the guise of free physical examinations. Three Chinese university and municipal partners cooperated in the study. The samples were exported to the US university's genetic bank for research into asthma (16,400 samples), diabetes, hypertension, and other diseases. Following complaints and media attention, leading to an international controversy, an investigation by the US Department of Health and Human Services was subsequently reported to have found serious violations in multiple respects, yet their published results stated in 2003 that as no participant had been harmed, no action would be taken. This was despite the study recruiting 16,686 asthma participants when only 2,000 had been approved, and taking larger volumes of blood than agreed. The approved amounts of financial compensation for travel and loss of work were also reduced from USD 10 per day to USD 1.50–3. It also emerged that many of the farmers had not provided consent as they were unaware this was a research study. Regulations in 2003 by the Chinese government to limit export of samples involving human genetic resources came too late to protect their samples. The US pharmaceutical company received major investment once it announced its possession of the samples. Several of the company's senior executives earned a net profit of over USD 10 million each through trade in stocks. The local residents who provided the samples received a free meal and an insignificant sum of money for expenses, many without knowledge of what was happening to them.

Equal partnerships with researchers from HICs are difficult to develop for countries like China. Loopholes and regulatory vacuums in host nations are easily exploited; it takes time to develop and introduce new systems and structures of research governance, but China's strengthening of its protection for Intellectual Property Rights (IPR), genetic resources and ethical review are now having a positive impact on the exploitation of its resources.[6]

International Genomics Research Involving the San People

In 2010, an international genomic research project entitled "Complete Khoisan and Bantu Genomes from southern Africa" was published in *Nature* amidst wide publicity (Schuster et al., 2010). The research aimed to examine the genetic structure of "indigenous hunter-gatherer peoples" selected from Namibia, and to compare the results with "Bantu from Southern Africa" (Schuster et al., 2010), including Nobel Peace Prize winner Archbishop Desmond Tutu. Four illiterate San elders were chosen for genome sequencing, and the published article analyzed many aspects of the correlations, differences, and relationships found in the single-nucleotide polymorphisms. A supplementary document published with the paper contained numerous conclusions and details that the San regarded as private, pejorative, discriminatory, and inappropriate. The use of sensitive and problematic terms (such as "Bushmen" and "hunter-gatherers") demonstrated a lack of awareness and consultation, while discussion of marriage and other cultural practices, speculation about lactase persistence in adults, the claim that "Bushmen have better hearing than Europeans," and the selective survival advantages of different levels of skin pigmentation, was deeply problematic for the San people. The San leadership (via the Working Group of Indigenous Minorities in Southern Africa [WIMSA]) met with the authors in Namibia soon after publication. They enquired about the informed consent process and asked why San leaders (via any one of three legitimated representative advocacy groups) had not been approached for permission in advance in accordance with international guidelines regarding research with indigenous peoples, particularly as genomic research by its very nature speaks to collective issues. The authors refused to provide details, apart from stating that they had received video-recorded consents in each case (Chennells & Steenkamp, 2018). They defended their denial of the right of the San leadership to further information on the grounds that the research project had been fully approved by ethics committees/institutional review boards in three countries tasked with "protecting the rights and welfare of research subjects" and that they had complied with all relevant requirements to respect the "culture, dignity and wishes of subjects" (Chennells & Steenkamp, 2018). The San leadership eventually wrote to *Nature*, expressing their anger at the inherent insult and lack of respect displayed by the whole process.[7]

Seeking Retrospective Approval for a Study in Resource-constrained Liberia

In 2014, the Ebola Virus Disease (EVD) epidemic hit Liberia very hard. At the peak of the epidemic in October 2014, researchers were arriving in the country to conduct all forms of research, including social science, anthropological, and clinical studies. As a result of the health emergency, the fully functional ethical research and oversight institutions were overwhelmed with investigators seeking information and guidance about the review process. Some investigators proceeded to conduct studies without the approval of institutional review boards (IRBs). At the height of the EVD surge (November–December 2014) one study gathered information on the economic well-being of EVD survivors in several communities to assess the economic impact of stigma and discrimination. Most of the participants were, at best, semi-literate. This study was therefore potentially distressing and traumatizing in itself and needed careful ethical consideration. However, the IRB did not receive an application until *after* the research was complete, when a junior research assistant representing the lead researcher (who had by then left the country) at a meeting revealed that approval was only being sought in order to disseminate the results. The IRB ruled that the research had been conducted unethically in contravention of clear national regulations and that the participants' autonomy had been breached. Approval was not given retrospectively, in a decision that took a stance on public policy and increasing compliance with mandated procedures.[8]

TRUST'S GLOBAL ENGAGEMENT ACTIVITIES

It is an imperative for TRUST's developments that the voices and input of all relevant stakeholder groups are heard. From a "top-down" perspective, this included a Funder Platform to engage members of research funding organizations around the world, who were brought together in a workshop in London in June 2017 to contribute to the development of the three TRUST tools. Funders are in a powerful position to demand adherence to high ethical standards, and to advise on their appropriateness, both ethically and in terms of compliance. Likewise, an industry platform was established to create a network of health sector industry representatives who share the vision of inclusive and fair research. Feedback from this network has informed TRUST's work, identified areas of (potential) exploitation in international research, and helped us to identify the necessary conditions by which industry might be willing to adopt the TRUST model for equitable partnerships in research worldwide.

In addition to the top-down input from a wide range of experts (Fig. 1), TRUST sought to actively bring forth the experiences and opinions of vulnerable research populations. It can be challenging to reach relevant stakeholders when consulting in resource-poor settings. Hence, to ensure appropriate representation, TRUST included two highly vulnerable populations in the project, through Partners for Health and Development in Africa (PHDA), representing the interests of Nairobi sex workers, and the South African San Institute (SASI), representing the San people of South Africa. Participation of these two groups

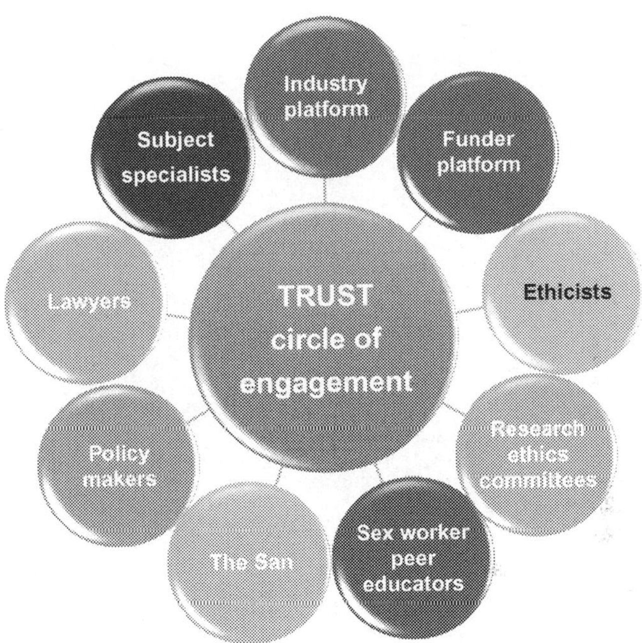

Fig. 1. Engagement Activities Brought Together a Broad Range of Stakeholders Who Worked Collectively to Achieve the TRUST Goals.

has ensured influential input from some of the most vulnerable research participants in the world.

Stakeholder engagement has been ongoing throughout the project via all of the typical channels, including email and online discussions, newsletters, and social media. In addition, specific events were organized to engage with stakeholders in an in-depth, face-to-face manner, as mapped on the TRUST timeline (Fig. 2).

Some of these events have been described above. Findings from others are captured below to illustrate the breadth of the subject areas and the richness of the information that was shared.

Engagement with the San and the Peer Educators for Sex Workers

Three workshops were held in Kimberley, South Africa, where representatives from the San community came together to consider their past experiences in research and how they might improve ethical standards for the future. Indigenous communities in all parts of the world can be vulnerable to intrusive research and exploitation of their knowledge with little or no benefit to themselves. The San, being one of the most highly researched populations globally, have many such exploitative experiences to recount (Wynberg, Schroeder, & Chennells, 2009). As a geneticist presenting at the first Kimberley meeting

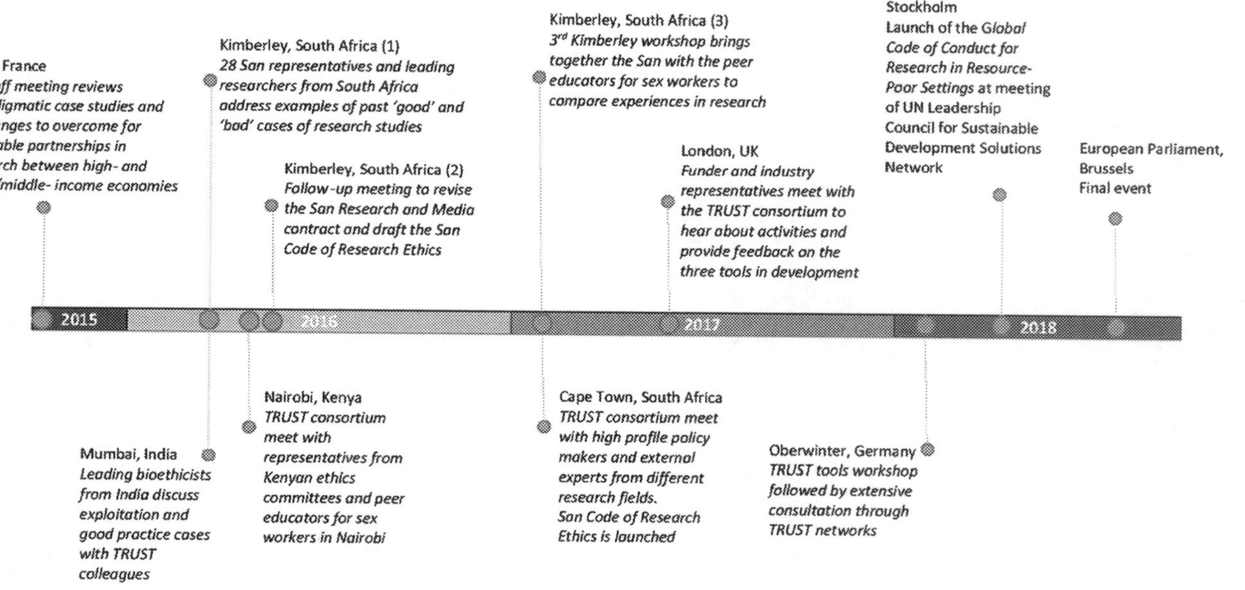

Fig. 2. The Trust Project Timeline (2015–2018).

explained, genetic research with the San is highly valued because they have the oldest lineage of all living populations on earth. Additionally, their traditional knowledge and their customs can be of great interest; a linguist at the same meeting described the importance of threatened languages, like those of the remaining 12 (out of a known 35) San languages.

Findings from the first workshop highlighted the following as major concerns for the San:

- a lack of appropriate informed consent procedures,
- the risks of not knowing or understanding the potential consequences of research,
- holding unrealistic expectations of participation,
- the vulnerability of individual participants, and
- a lack of negotiation with community leaders and mandated organizations to avoid confusion and conflict, and ensure respect for cultural requirements.

However, the San also described experiences of good practice where there has been clarity of intention of the researchers, effective informed consent strategies, respect for local research governance and negotiation with community representatives, a clear negotiation of benefits for research participants/communities, and the delivery of results and promised outcomes.

Looking forward, a need for capacity building was identified, especially pertaining to legal and contractual matters such as intellectual property rights, copyright law, and contracts. Other needs included consequences for those who do not abide by the existing San *Research and Media Contract*, which was felt to be in need of revision, and a formal code of research ethics for studies involving the San, which should be binding for researchers.

These themes were developed into outputs at the second consultative workshop in Kimberley, two months after the first. Here, 22 San representatives contributed to the drafting of the *San Code of Research Ethics* (South African San Institute, 2017), the first code of research ethics to be developed by an indigenous community in Africa. Additionally, revisions to the *Research and Media Contract* were suggested in line with the new Code and it was agreed that the San Council should be centrally responsible for research management.

The first major input from the Kenyan peer educators for sex workers came in Nairobi, May 2016. In Nairobi, PHDA runs the Sex Workers Outreach Programme, providing clinical and preventative services to sex workers who would otherwise find it difficult to access public health services because of discrimination and stigma. Those enrolled at the clinic for HIV prevention services are invited to participate in research studies concerning the epidemiology of sexually transmitted diseases and the host genetic factors that influence infectivity and disease progression (Andanda & Cook Lucas, 2007). Here, five representatives from the sex worker community spoke at length about their experiences of participating in clinical research studies and informed us about risk factors for their exploitation.[9] They expressed their community's concerns in the following areas:

- *Informed consent*
  Information needs to be fully accessible to those with low or no literacy, in appropriate languages, with clear and honest information about potential risks, including how those will be managed, and any benefits. Engaged communication is necessary as rumors can spread swiftly in close communities. Researchers must recognize that resource-poor people are at high risk of exploitation; most of them consent to participation because of the cash incentives and the possibility of health benefits.[10]
- *Feedback*
  The sex worker community needs feedback from research studies in simple and non-scientific language. In the past, results have been fed back in technical language that they can't understand: it puts people at risk when they don't understand the results.
- *Input into research design*
  There is a desire and willingness to participate fully in studies, right from the research design stage. Some of the community have suitable qualifications and this could help to broaden research literacy, as well as improve trust in the research and ease recruitment. Proper remuneration for such roles would be essential.
- *Specific sex worker concerns*
  Specific concerns such as mental health issues, addiction, and alcoholism need to be considered, as well as the needs of those with HIV. Sex work is illegal in Kenya, so there are always fears about confidentiality. There is more trust if researchers approach the community through the dedicated clinics, highlighting the importance of developing long-term relationships.[11]
- *Cultural sensitivity*
  Proper engagement with the community prior to the research is necessary to understand cultural sensitivities and take them into account. For example, there are concerns about the destinations of samples due to cultural beliefs, stigmatization of gay men, and general prejudice against sex workers.

In the third Kimberley meeting, the five peer educators from the Nairobi sex worker community joined with the San for a workshop to explore differences and similarities between their experiences. A dominant theme of this dialog was that there are many common concerns across the different settings. As in many situations, where there is a serious power imbalance, this demands fully engaged dialogue and comprehensive consent. Many of the most vulnerable populations are marginalized both culturally and legally, which can also create ambiguities in the concept and practice of leadership. Furthermore, researchers' models of community engagement need to be appropriate, rather than just exported from other settings.

Table 1. Primary Concerns and Challenges for Kenyan Research Ethics Committees Regarding International Collaborative Research.

Research Governance Challenges	Research Ethics Challenges
• Different governance standards and procedures	• Exploitation of local researchers
• Unwillingness of Northern partners to abide by double ethics review	• No or little local relevance, or research outputs not affordable
• Ethics dumping potential due to REC oversight, capacity and training problems resulting from resource constraints	• Northern-type informed consent procedures ignoring literacy levels and community consent
• Unresolved issues in the ownership of biological samples	• Northern researchers show no or little cultural sensitivity
• Unresolved issues in the ownership of primary data	• Lack of feedback/dissemination
	• Standards of care or placebo use differ between partners

A Kenyan Research Ethics Committee Perspective

In addition to the Kenyan participant and community perspective provided by the sex worker peer educators, the Nairobi meeting also heard an ethics committee and governance perspective from three of the most senior research ethics committee (REC) chairs in Kenya. These three professors provided direct insights into their considerable experience of the ethical challenges they have encountered when dealing with international, collaborative research. Together they revealed a broad range of challenges which allow for the potential exploitation of Kenyan research participants, Kenyan researchers, and Kenyan resources.

The identified concerns and challenges are summarized into two categories: those that are largely issues of research governance and those that are more obviously issues of research ethics. By "research governance" we mean the processes and systems that are used to ensure the regulation of research. For instance, one *process* or *system* by which the ethical acceptability of research can be achieved is double ethics review: approval from both the sponsor country and the local host. By "research ethics" we mean the moral requirements that guide the conduct of research. For instance, the special protection that has to be given to vulnerable populations in research is a substantial requirement that refers directly to moral principles (protecting those who cannot protect themselves). Table 1 illustrates the breadth of the challenges that were described.

The concerns and challenges related to research governance are primarily associated with a need for more resources and tighter legal and regulatory systems. Neither is within the control of RECs. The research ethics challenges and concerns for Kenya echo what other authors have observed across LMICs.[12]

A Meeting of Many Minds

In Cape Town, a plenary meeting in 2017 broadened developing themes with the inclusion of perspectives from other fields in collaborative research: agriculture and biodiversity, technology transfer, and animal research. It brought together experts from these fields with people who are well placed to influence funding agencies, national government departments and science councils, and the project partners, including representatives from the San and the Nairobi sex workers community. Attendees at this meeting were invited to consider and input directly into the emerging TRUST tools, providing invaluable insights that helped to steer subsequent developments. Contributions from these perspectives have been vital for ensuring that the tools are realistic and practicable across a range of research disciplines.

At the culmination of this meeting, representatives from various San communities in southern Africa came together with the TRUST team for the official launch of the *San Code of Research Ethics* (SASI, 2017). This Code has subsequently garnered much publicity and support, with enthusiastic interest from journalists, researchers, ethics committees, funders, and other vulnerable populations around the world.[13]

THE FOUR TRUST VALUES

One of the outcomes of the Nairobi event was agreement that a set of values should guide the development of the Global Code of Conduct as well as TRUST's other outputs. The current international ethics framework is heavily influenced by an applied ethics approach that originated in the United States. This approach concerns the moral permissibility of specific actions and refers primarily to four principles (autonomy, beneficence, non-maleficence, and justice) that represent the cornerstone of biomedical ethics (Beauchamp & Childress, 2009). These principles are widely applied beyond their origins in biomedical research, with varying degrees of acceptance and applicability. Through our engagement activities, it has become clear that these four principles have problems with global applicability and common global understanding.[14] Hence, rather than simply adopting this existing ethical framework, we have used an alternative approach that we believe resonates across borders and cultural contexts.

In TRUST we refer to values, rather than principles, as the foundation of our ethical standpoint in research. Values can be understood as the beliefs people have, especially about what is right and wrong and most important in life, that influence their behavior. As such, they can inspire, motivate, and engage people to discharge obligations or duties. The four TRUST values have been agreed as Fairness, Respect, Care, and Honesty:

Fairness

Fairness (or justice) can have a number of interpretations, but the most relevant concepts for collaborative research ethics are *fairness in exchange* and *corrective fairness*. In collaborations at least two parties are involved in a range of transactions, and issues that need to be considered for fairness in exchange might include the opportunities and allocation of benefits from the research for all parties. Corrective fairness is about how to right a wrong and includes considerations such as liability and accountability. This type of fairness is vital in collaborative ventures but can be challenging because it is dependent upon the availability and applicability of legal instruments and access to mechanisms to right a wrong (e.g., a complaints procedure, a court, an ethics committee).

Respect

To show respect when engaging with communities requires an acceptance that their customs and cultures may be different from your own, and that you should behave in a way that does not cause offense. It means that one may need to accept a decision or a way of approaching a matter, even if one disagrees. Respect is therefore also a difficult value, as there is always the possibility that one *cannot* accept another's decision, especially when this creates a serious conflict of conscience. To find an appropriate route between imperialist-type imposition of approaches and careless acceptance of human rights violations may sometimes be challenging, but it is what researchers with integrity must sometimes address. And if they cannot address it, it may not be possible to undertake the research.

Care

As a priority, care should be taken of those enrolled in research studies to the extent that their welfare is prioritized over any other goals. In line with the *Declaration of Helsinki* this means: "While the primary purpose of medical research is to generate new knowledge, this goal can never take precedence over the rights and interests of individual research subjects" (World Medical Association [WMA], 2013, Article 8).

This care value applies across disciplines, not just in medical research and is also not restricted to human research participants. Article 21 of the *Declaration of Helsinki* extends the care for welfare to research animals (WMA, 2013, Art. 21). Likewise, care for environmental protection and sustainability is increasingly included in research ethics processes and frameworks for responsible research (Owen, Macnaghten, & Stilgoe, 2012).

Researchers who take good care combine two elements: they care about research participants, in the sense that they are important to them, *and* they feel responsible for the welfare of those who contribute to their research, or might suffer as a result of it (including animals and the environment).

Honesty

Honesty is a value that does not need complicated explanations or definitions. In all cultures and nations, "do not lie" is a basic prerequisite for ethical human interaction. However, what does need explaining is the scope of the value of honesty in the context of global research ethics.

Lying is only one possible wrongdoing in the context of a broad understanding of honesty. For instance, in research ethics it is equally unacceptable to omit important information from an informed consent process. For this reason, research ethicists often use the terms *transparency* or *open communication* to ensure that all relevant information is provided so that research participants can make an informed choice about participating or not. Importantly, honesty is also related to research conduct other than interaction with research participants. Most prominently, the duties of honesty are described in *research integrity* frameworks (which are increasingly binding on researchers in institutions and via funders' requirements), which include issues such as credit for contributions, manipulation of data, or misappropriation of research funds (World Conference on Research Integrity, 2010).

DEVELOPING A GLOBAL CODE OF CONDUCT FOR RESEARCH IN RESOURCE-POOR SETTINGS

Developing a Code of Conduct that has global applicability is no easy task. We have been mindful from the outset that there were specific traps to be avoided:

- *Trying to reinvent the wheel.* There are a multitude of research ethics codes already in existence and it would be foolish to ignore them.
- *Building a conglomerate of existing codes.* While they may contain some relevant elements, most of the existing codes have been authored by people in high-income settings and are not directly focused upon the challenges that are associated with collaborative ventures in poor-resource settings.
- *Producing a "we know it all" code.* Specialist expertise can be extremely helpful but may not capture all relevant vulnerabilities for ethics dumping. We wanted a systematic grounding for our code.

Before we could even begin to imagine what a code might look like, it was vital for us to understand what makes exploitation more likely to occur due to vulnerabilities that can be exploited, either knowingly or unknowingly. Investigation of this vast subject would be impossible from a traditional literature-based approach, or through investigation in a single geographical region. Many of these vulnerabilities are poorly represented in the literature and they can differ between countries, cultures, and the nature of the research. For example, clinical trials, social science, animal experiments, environmental science, and research in emergency settings may pose a diverse array of risks that are largely dependent upon the local context in which they are undertaken. A creative approach to data collection was needed to try and capture as many risks and vulnerabilities as possible, hence our emphasis on wide-ranging stakeholder engagement. Our Code is rooted in a broad-based consultative exercise incorporating input from all of the aforementioned engagement activities.[15]

Individual vulnerabilities and risks of exploitation were extracted from the vast amount of information provided and the raw data were collapsed to group similar vulnerabilities together. For instance, there were many different examples of how people living in poor circumstances may be unfairly enticed to participate in research by the prospect of payment or reward. Such examples were grouped under the label of "undue inducement." Further thematic analysis resulted in distinctions between the various potential subjects, or levels of risk for exploitation (persons, institutions, local communities, countries, animals, and the environment); in the final stage of the analysis, the vulnerabilities were grouped according to the four values of fairness, respect, care, and honesty.

This exercise resulted in an Exploitation Risk Table (Chatfield et al., 2016) that contained 88 risks for exploitation in collaborative research. Importantly, care was taken to ensure that each individual risk was based upon real-world experience rather than hypothetical suppositions. The exploitation risk table clearly highlights the vulnerabilities that need to be taken into consideration when working in resource-poor settings in order to avoid ethics dumping. When risks were mapped against existing codes for research ethics, it was found that most (79) were addressed, at least to some extent, by an element in an existing code. However, no existing code addressed them all. Furthermore, it is not easy to spot the elements in existing codes that are of special significance to collaborative research in LMICs unless one is already aware of the challenges.

Our Global Code of Conduct for Research in Resource-Poor Settings consists of 23 articles, grouped according to the four values. Collectively, they address the 88 risks for exploitation that we identified. For example, under the Fairness value, Article 1 addresses risks to communities and institutions, as shown in Table 2.

The Code does not repeat standard requirements for ethical research that apply wherever researchers work, across all settings. It lists only those that apply

Table 2. Article 1: Global Code of Conduct for Research in Resource-poor Settings Addresses Issues of Fairness that Affect Communities and Institutions.

Article 1	Addresses Risks of Exploitation For:
Local relevance of research is essential and should be determined in collaboration with local partners. Research that is not relevant in the location where it is undertaken imposes burdens without benefits	*Communities:* LMIC communities can be exploited in research when aims are driven by, and in the interests of, high-income researchers/institutions with no real benefit to the local community. If the research is of no potential benefit to the local community, we must ask why it is being conducted there. *Institutions:* Where LMIC partners are dependent upon funding and association with their high-income partners for research, the research aims may be shaped by the high-income partners and not tailored to the preferences, needs, and skills of the local workforce.

when people from high-income settings are working in poor-resource settings. In this way, it offers a straightforward, quick, and user-friendly means of ascertaining the ethical requirements for collaborative ventures with LMICs.

In summary, the Global Code of Conduct counters ethics dumping by:

- providing guidance across all research disciplines,
- presenting clear, short statements in simple language to achieve the highest possible accessibility,
- focusing on research collaborations that entail considerable imbalances of power, resources, and knowledge, and
- using a new framework based on the values of fairness, respect, care, and honesty.

To ensure impact and longevity of the Code and going beyond the lifetime of the project, a stand-alone website was created[16] that includes considerable learning materials in a resource hub, to support the Code.

The Code was launched in May 2018 at a meeting of the UN Leadership Council of the Sustainable Development Solutions Network, one of the most influential groups internationally working on global justice issues. One month later, in June 2018, it was distributed at a European Parliament event to members of parliaments, journalists, academics, and the general public. In his speech, Wolfgang Burtscher, the European Commission's Deputy Director-General for Research announced:

> As a concrete step forward, I would like to inform you that the Code developed by TRUST will be soon included in the Participant Portal of Horizon 2020 as a reference document to be consulted and applied by all relevant research projects and serve as an education tool for the younger generation of researchers.

The TRUST group was inspired by the multi-stakeholder approach to code building and its result. We hope the Code will inspire researchers to build equitable research relationships between HICs and LMICs so that the benefits of innovative research will become available to all.

NOTES

1. The term was first used by the Science with and for Society Unit of the European Commission, which defines it as follows: "Due to the progressive globalisation of research activities, the risk is higher that research with sensitive ethical issues is conducted by European organisations outside the EU in a way that would not be accepted in Europe from an ethical point of view. This exportation of these non-compliant research practices is called ethics dumping" (European Commission, 2016).

2. *Creating and enhancing TRUSTworthy, responsible and equitable partnerships in international research* is a 3-year (2015–18) project funded by the European Union's Horizon 2020 research and innovation programme, grant agreement No 664771. Retrieved from http://trust-project.eu/the-project/about/. Accessed on May 23, 2018.

3. Where households cannot meet basic needs for survival (e.g., chronic hunger, no access to health care).

4. Where households can only just meet basic needs for survival, with little left for the education of their children.

5. Taken from Srinivasan, Johari, and Jesani (2018).
6. Taken from Zhao and Yang (2018).
7. Taken from Chennells and Steenkamp (2018).
8. Taken from Tegli (2018).
9. Sex workers in LMICs are among the most vulnerable and frequently researched populations. Markers for their extreme vulnerability are the fact that they are 14 times more likely to contract HIV compared to other citizens in their countries and that they carry a very high burden of violence. Retrieved from http://www.jhsph.edu/news/news-releases/2012/baral-sex-workers.html. Accessed on May 23, 2018] and Retrieved from http://www.ncbi.nlm.nih.gov/pubmed/24625169. Aaccessed on May 23, 2018.
10. See Cook Lucas et al. (2013).
11. See Tukai (2018).
12. See, for instance, Joseph, Caldwell, Tong, Hanson, and Craig (2016).
13. Retrieved from http://trust-project.eu/san-code-of-research-ethics/
14. For instance, often viewed as a liberalist ideal, Beauchamp & Childress' (2009) interpretation of "autonomy" focuses on the rights of individuals to choose what happens to their bodies. However, this view can be at odds with cultural norms and practices in environments where the well-being of the community as a whole is more highly valued than that of any individual.
15. This type of consultative exercise is of proven value in the development of ethical codes that are broadly representative and can have wide-ranging impact; the principles of the "Three Rs," which are globally accepted as a reasonable measure for ethical conduct in animal research, arose from a broad consultation with stakeholders undertaken in the 1950s. See Russell, Burch, and Hume (1959).
16. Retrieved from http://www.globalcodeofconduct.org/

REFERENCES

Andanda, P., & Cook Lucas, J. (2007). *Majengo HIV/AIDS research case: a report for GenBenefit*. Retrieved from http://www.uclan.ac.uk/research/explore/projects/assets/cpe_genbenefit_nairobi_case.pdf

Beauchamp, T. L., & Childress, J. F. (2009). *Principles of Biomedical Ethics* (6th ed.), New York, NY: Oxford University Press.

Bradley, M. (2017, February 15). Whose agenda? Power, policies, and priorities in North–South research partnerships. In L. Mougeot (Ed.), *Putting knowledge to work: Collaborating, influencing and learning for international development* (pp. 37–70): Practical Action Publishing, IDRC. Retrieved from https://www.idrc.ca/en/book/putting-knowledge-work

Chatfield, K., Schroeder, D., Leisinger, K., van Niekerk, J., Munuo, N., Wynberg, R., & Woodgate, P. (2016). *Generic risks of exporting non-ethical practices a report for TRUST*. Retrieved from http://trust-project.eu/wp-content/uploads/2016/12/TRUST-Deliverable-Generic-Risks-Final-copy.pdf

Chennells, R., & Steenkamp, A. (2018). International genomics research Involving the San people. In D. Schroeder, J. Cook, F. Hirsch, S. Fenet, & V. Muthuswamy (Eds.), *Ethics dumping, Case studies from North-South research collaborations* (pp. 15–22). New York, NY: Springer.

Cook Lucas, J., Schroeder, D., Arnason, G., Andanda, P., Kimani, J., Fournier, V., & Krishamurthy, M. (2013). Donating human samples: Who benefits? Cases from Iceland, Kenya and Indonesia. In D. Schroeder & J. Cook Lucas (Eds.), *Benefit sharing: From biodiversity to human genetics* (pp. 95–128). Dordrecht: Springer.

Dickson, D. (2006). Calling into question clinical trials in developing countries. *The Lancet*, *368*(9549), 1761–1762.

European Commission. (2010). *European textbook on ethics in research*. Luxembourg: Publications Office of the European Union. Retreived from https://ec.europa.eu/research/science-society/document_library/pdf_06/textbook-on-ethics-report_en.pdf

European Commission. (2016). *Horizon 2020, the EU framework programme for research and innovation: Ethics 2016*. Retrieved from https://ec.europa.eu/programmes/horizon2020/en/h2020-section/ethics. Accessed on September 19, 2018.

Gainotti, S., Turner, C., Woods, S., Kole, A., McCormack, P., Lochmüller, H., & Taruscio, D. (2016). Improving the informed consent process in international collaborative rare disease research: Effective consent for effective research. *European Journal of Human Genetics*, *24*(9), 1248.

Glickman, S. W., McHutchison, J. G., Peterson, E. D., Cairns, C. B., Harrington, R. A., Califf, R. M., & Schulman, K. A. (2009, Feb 19). Ethical and scientific implications of the globalization of clinical research. *New England Journal of Medicine*, *360*(8), 816–823.

Investopedia. (2018). What is dumping? Retrieved from https://www.investopedia.com/terms/d/dumping.asp

Joseph, P. D., Caldwell, P. H., Tong, A., Hanson, C. S., & Craig, J. C. (2016, Feb). Stakeholder views of clinical trials in Low-and Middle-Income Countries: A Systematic Review. *Pediatrics*, *137*(2), e20152800. doi:10.1542/peds.2015-2800.

Luna, F. (2009). Research in developing countries. In B. Steinbock (Ed.), *The Oxford handbook of bioethics* (pp. 621–647). Oxford: Oxford University Press. Part VII. Chapter 26. Retrieved from http://www.oxfordhandbooks.com/view/10.1093/oxfordhb/9780199562411.001.0001/oxfordhb-9780199562411-e-027

Macklin, R. (2004). *Double standards in medical research in developing countries*. Cambridge: Cambridge University Press.

Molyneux, C., Goudge, J., Russell, S., Chuma, J., Gumede, T., & Gilson, L. (2009). Conducting health-related social science research in low income settings: Ethical dilemmas faced in Kenya and South Africa. *Journal of International Development*, *21*(2), 309–326.

Novoa-Heckel, G., Bernabe, R., & Linares, J. (2017). Exportation of unethical practices to low and middle income countries in biomedical research. *Revista de Bioética y Derecho*, (40), 167–179.

Owen, R., Macnaghten, P., & Stilgoe, J. (2012). Responsible research and innovation: From science in society to science for society, with society. *Science and public policy*, *39*(6), 751–760.

Ravinetto, R., Tinto, H., Diro, E., Okebe, J., Mahendradhata, Y., Rijal, S., & De Nys, K. (2016). It is time to revise the international Good Clinical Practices guidelines: Recommendations from non-commercial North–South collaborative trials. *BMJ Global Health*, *1*(3), e000122.

Russell, W. M. S., Burch, R. L., & Hume, C. W. (1959). *The principles of humane experimental technique*. London: Methuen & Co.

Schroeder, D., Cook, J., Hirsch, F., Fenet, S., & Muthaswamy, V. (Eds.). (2018). *Ethics dumping – Case studies from North-South research collaborations*. New York, NY: Springer.

Schroeder, D., & Gefenas, E. (2009). Vulnerability – Too vague and too broad? *Cambridge Quarterly of Healthcare Ethics*, *18*(2), 113–121.

Schuster, S. C., Miller, W., Ratan, A., Tomsho, L. P., Giardine, B., Kasson, L. R., … Hayes, V. M. (2010). Complete Khoisan and Bantu genomes from southern Africa. *Nature*, *463*, 943–947. Retrieved from http://dx.doi.org/10.1038/nature08795.

South African San Institute (SASI). (2017). *San Code of Research Ethics*. Retrieved from http://trust-project.eu/wp-content/uploads/2017/03/San-Code-of-RESEARCH-Ethics-Booklet-final.pdf

Srinivasan, S., Johari, V., & Jesani, A. (2018). Cervical cancer screening in India. In D. Schroeder, J. Cook, F. Hirsch, S. Fenet, & V. Muthuswamy (Eds.), *Ethics dumping - Case studies from North-South research collaborations* (pp. 33–48). New York, NY: Springer.

Tegli, J. K. (2018). Seeking retrospective approval for a study in resource-constrained Liberia. In D. Schroeder, J. Cook, F. Hirsch, S. Fenet, & V. Muthuswamy (Eds.), *Ethics dumping - Case studies from North-South research collaborations* (pp. 115–120). New York, NY: Springer.

Tukai, A. (2018). Sex workers involved in HIV/AIDS research. In D. Schroeder, J. Cook, F. Hirsch, S. Fenet, & V. Muthuswamy (Eds.), *Ethics dumping – Case studies from North-South research collaborations* (pp. 23–31). New York, NY: Springer.

World Conference on Research Integrity. (2010). *Singapore statement on research integrity*. Retrieved from http://www.singaporestatement.org/

World Medical Association (WMA). (2013). *Declaration of Helsinki – Ethical principles for medical research involving human subjects*. Retrieved from https://www.wma.net/policies-post/wma-declaration-of-helsinki-ethical-principles-for-medical-research-involving-human-subjects/

Wynberg, R., Schroeder, D., & Chennells, R. (2009). *Indigenous peoples, consent and benefit sharing*. Dordrecht: Springer.

Zhao, Y., & Yang, W. (2018). An international collaborative genetic research project conducted in China. In D. Schroeder, J. Cook, F. Hirsch, S. Fenet, & V. Muthuswamy (Eds.), *Ethics dumping – Case studies from North-South research collaborations* (pp. 71–80). New York, NY: Springer.

EBOLA VIRUS DISEASE: A LESSON IN SCIENCE AND ETHICS

Nicola Petrosillo and Rok Čivljak

ABSTRACT

The largest ever outbreak of Ebola virus disease (EVD), which began in December 2013, profoundly impacted not only the West African countries of Guinea, Sierra Leone, and Liberia, and to a lesser extent Nigeria, but also the rest of the world because some patients needed to be managed in high-resource countries. As of March 29, 2016, there were 28,616 confirmed, probable, and suspected cases of EVD reported in Guinea, Liberia, and Sierra Leone during the outbreak, with 11,310 deaths (case fatality rate of 39.5%). An unprecedented number of healthcare workers and professionals, including physicians, nurses, logistic and administrative personnel, housekeepers, epidemiologists, statisticians, psychologists, sociologists, and ethics experts in many countries, were directly or indirectly involved in the care of EVD patients.

The provision of medical care to critically ill EVD patients would have been challenging in any setting but was especially so in the remote and resource-limited areas where patients were stricken by EVD. Limited health personnel, medical supplies, and equipment, along with inadequate knowledge and skills for minimizing the risks of transmission to healthcare workers, could have led to the de-prioritization of patient care. However, ethical considerations demanded aggressive patient care (intensive care, dialysis, central vascular catheter indwelling, etc.) to produce positive outcomes without increasing the risks to healthcare workers and caregivers.

A major ethical consideration was that healthcare workers have a double obligation: while providing the best medical care to improve EVD patient survival, with symptom relief and palliation as required, they must also protect themselves and minimize further transmission to others, including their colleagues. During the 2014–2015 EVD epidemic, another ethical and

Ethics and Integrity in Health and Life Sciences Research
Advances in Research Ethics and Integrity, Volume 4, 33–44

ISSN: 2398-6018/doi:10.1108/S2398-601820180000004003

clinical problem arose in relation to the management of healthcare workers deployed in Africa who acquired EVD while caring for infected patients. As of June 24, 2015, a total of 65 individuals had been evacuated or repatriated worldwide from the EVD-affected countries, of whom 38 individuals were evacuated or repatriated to Europe. The need for evacuation and repatriation, together with associated ethical issues, is discussed in this chapter.

Keywords: Ebola; ethics; healthcare workers; outbreak; medical care; resource-limited countries

EBOLA: A PERSONAL EXPERIENCE IN LAGOS, NIGERIA

In August 2014, I was deployed as a World Health Organization (WHO) clinical consultant to the Ebola virus disease (EVD) response team in Lagos, Nigeria. News reports coming from West African countries experiencing dramatic increases in EVD cases were all over the front pages of newspapers, with distressing photos and stories. Moreover, a high percentage of EVD cases in Nigeria involved healthcare workers. To be honest, I was a bit worried. Before leaving my country, I was given several vaccines (yellow fever, hepatitis, meningitis etc.), which made me very nervous. Although keenly aware of the risks, I chose to help despite them.

My tasks were to provide technical advice on clinical care to the EVD Care Center at the Mainland Hospital in Yaba, a district of Lagos, to augment infection prevention and control practice training and provide technical advice for other activities by the Nigerian response team. Nigeria had experienced a successfully contained EVD outbreak, introduced by a Liberian-American lawyer incubating EVD, who travelled from Liberia to Lagos, denied previous exposure to EVD after he became sick and infected several healthcare workers who cared for him.

Since the first day in Lagos, I was aware that it would take a day or so to become oriented and to understand where I was. Lagos is a big city with 21 million inhabitants, many of whom are on the streets daily. Bush meat is everywhere and the traffic is terrible. Before leaving for Lagos, I was told not to walk alone and to be very careful when travelling on the roads. As soon as I arrived in Lagos, I understood the importance of following these safety precautions.

The isolation facility for suspected/confirmed EVD cases was in a compound for tuberculosis, which is endemic in Nigeria. We met the local healthcare workers in a large empty ward with several beds, which had previously been for pediatric tuberculosis patients, whose relatives looked after them there. I dare not imagine how respiratory precautions had been applied in that ward. Nevertheless, I realized that after the Ebola crisis passes, tuberculosis will still remain.

The isolation facility was located in the back of a compound with three wards, one connected to its own entrance gate that could accommodate 8–10 isolation beds, one inside with 14 beds for suspected EVD cases, and another with 14 beds for lab-confirmed EVD patients. I was replacing a WHO physician with extensive experience in EVD. Médecins Sans Frontières/Doctors Without Borders (MSF) volunteers had joined the response team a few days after his arrival, with an outbreak assessment and support team that included a team lead (clinician) with 2 years of EVD outbreak care experience, and a logistics person familiar with the usual water/sanitation issues. Logistics is crucial in the struggle against Ebola.

Thanks to the coordination of the MSF/WHO groups, provision of personal protective equipment (PPE) was guaranteed, and the logistics/water sanitation equipment in the facility ensured a safe work environment for clinicians, nurses and housekeepers. Nigerian participation in the healthcare of suspected/confirmed EVD patients was increasing; some Nigerian doctors and nurses were staffing the care center and others were being trained. Besides epidemiological, infection prevention and control, and clinical training, a scheduled observation shift and then three scheduled shadow shifts were expected. The trainees were all young, willing and friendly, but their presence was unpredictable. The work was intense and problems appeared everywhere. The medical team shared a large room, where they had no opportunity to rest or conduct focused meetings. There were people bustling around all day long. Sometimes it was preferable to hold briefings in the open-air garden in front of the isolation facility. Another problem was the shortage of local personnel to cover three shifts, which was partly overcome by the enthusiastic young staffers. The local healthcare workers were conscious of the extraordinary nature of the event in which they were participating.

My daily duties included attending a large meeting with representatives of the Nigerian Ministry of Health, local representatives, epidemiologists, logistics personnel etc., in a building that was a 40-minutes drive from the hospital. Then, we would drive to the EVD isolation facility through extremely heavy traffic on a road with few traffic lights. The day ended with another drive through traffic to an evening meeting, where all the teams (clinical, epidemiological, logistics etc.) reported on their activities (response team activities). Although the majority of the people attending this meeting were tired after the day's work, everyone paid close attention to each update, including the overall number of suspected and confirmed cases, contacts to be traced or already traced, and rumors of likely cases. Strategic issues, including points of entry measures, contact tracing, logistics, media and social networks, were also discussed by national and international experts from WHO, MSF, the United Nations Children's Fund (UNICEF), Centers for Disease Control and Prevention (CDC) etc.

The daily work was exhausting but not boring. Caring for suspected/confirmed EVD cases was exciting. The complex procedure for donning personal protection equipment, the heat, humidity and sweating, the constant disinfection with chlorine, the difficulty I experienced in approaching patients due to the protective suit that covered every part of my body, the poor diagnostic tools, and the dangerous and difficult procedure for doffing personal protective equipment all contributed to making the care of patients a challenge, and sometimes frustrating. Donning and doffing personal protective equipment were lengthy and painstaking activities. All the staff members were aware of the vital and life-saving importance of avoiding any contamination and infection from Ebola virus.

However, all our frustrations disappeared and were fully rewarded when a previously very sick patient improved and was discharged, such as a woman who, upon exiting from the dedicated passage of the isolation facility to the outside world, where her kids were waiting for her, cried "Thank God" and smiled at us in gratitude.

My personal experience was a life lesson. For the fight against EVD in Nigeria, ample financial and material resources, as well as well-trained and experienced international and national staff, were provided. Through efficient organization, with a strong coordinated commitment and international collaboration, the collective efforts of normal people acting with great humanity and self-sacrifice were successful. For Nigeria, the fight against EVD was a spectacular success story, which shows that Ebola can be contained.

Nicola Petrosillo

BACKGROUND

Ebola virus disease (EVD), formerly known as Ebola hemorrhagic fever, is a severe and often fatal illness in humans. It is caused by the Ebola virus (EBOV), which is transmitted to people from wild animals and spreads among the human population through human-to-human transmission. The average EVD case fatality rate is around 50%, ranging from 25% to 90% in past outbreaks.

The 2014–2016 Ebola outbreak in West Africa was the largest and most complex since the virus was first discovered in 1976. There were more cases and deaths during that outbreak than from all others combined. It also spread among countries, starting in Guinea and then moving across land borders to Sierra Leone and Liberia. From the beginning of the epidemic until it ended in April 2016, a total of 28,616 confirmed, probable, and suspected EVD cases were reported with 11,310 deaths (case fatality rate of 39.5%). Besides Guinea, Liberia, and Sierra Leone, other countries were involved in the epidemic, including Nigeria (20 cases and 8 deaths), Mali (8 cases and 6 deaths), and Senegal (one case, no deaths) (https://www.cdc.gov/vhf/ebola/outbreaks/2014-west-africa/index.html). Moreover, only a few EVD cases were evacuated from West Africa to Western countries. A total of 26 cases, of which 4 were fatal (Center for Infectious Diseases Research and Policy (CIDRAP)), were treated in Europe (Kreuels et al., 2014; Mora-Rillo et al., 2015; Wolf et al., 2015) and North America (Liddell et al., 2015; Lyon et al., 2014) during the outbreak. The Ebola virus disease outbreak reminded the world of the dangers of disease transfer from animal reservoirs,that is, zoonosis. An unprecedented number of healthcare professionals, including physicians, nurses, logistics and administrative personnel, housekeepers, epidemiologists, statisticians, psychologists, sociologists, ethics experts, etc. from a variety of clinical settings in a number of countries were directly or indirectly involved in caring for EVD patients. Guidance documents on infection prevention and control practice and clinical care were provided by organizations with EVD experience (Centers for Disease Control & Prevention CDC, 2014; Médecins Sans Frontières MSF, 2014; World Health Organization WHO, 2014).

As important as guidance documents are, many lessons must be learned from the specific hands-on daily care of EVD patients in countries with limited healthcare facilities, which can help us avoid some of the risks inherent in the steep learning curve associated with delivering EVD care.

CRITICAL POINTS IN THE CARE OF EVD PATIENTS (ESPECIALLY IN RESOURCE-LIMITED SETTINGS)

Ethically speaking, healthcare workers have a double obligation: to provide the best medical care in order to improve patient survival, with symptom relief and palliation as required, while at the same time they must also protect themselves and minimize further transmission to others, including their colleagues.

Anxiety, haste, and pressure, especially from politicians and the media, often blur the facts. The care of patients with EVD must be deliberate, vigilant, and guided by science. The basic procedures for how clinicians can safely approach a suspected/confirmed EVD patient, , which are based on decades of research and field observation, should be followed. Although much remains to be discovered, Ebola virus is only spread during the symptomatic phase of the illness, especially in the presence of diarrhea, vomiting, or bleeding, and we know the relevant epidemiological features, such as the incubation period, infected body fluids, and mechanisms of transmission. Understanding these principles helps eliminate the sense of mystery, reduces stress, and keeps responders focused on their work (Brett-Major et al., 2015).

Secondly, safe and effective care for EVD patients has been achieved in both resource-poor and well-resourced settings. The healthcare workers caring for a suspected/confirmed EVD patient should apply a targeted strategy for any of the clinical manifestations. Volume repletion and electrolyte management, attention to hypoperfusion-related complications, intravenous indwelling, hemodialysis in the event of acute renal insufficiency, and reanimation procedures, including mechanical ventilation, should be provided to patients when needed. Patient safety in the isolation environment and her/his psychological/mental status should be vigilantly managed. Differential diagnosis and the search for coinfections (malaria is endemic in many countries affected by Ebola) often require healthcare workers to perform blood-drawing and invasive procedures, although they risk contracting EVD from their patients unless rigorous contact and droplet and airborne safety precautions are scrupulously followed.

The highest mortality rate at the beginning of EVD epidemics may reflect the relatively low-level care EVD patients receive, owing to fear of viral transmission (healthcare workers' fear of exposure and infection). Later on, when many invasive procedures (peripheral and central venous access, dialysis, and mechanical ventilation) were being performed safely in the proper settings, mortality dramatically decreased. "First do no harm" applies to the patient, the staff, and the community (Brett-Major et al., 2015).

Another important point is represented by the attention of clinicians and all the healthcare staff to the safety of the environment in healthcare facilities. Indeed, the environmental aspects of wards/units are typically not managed by clinicians. However, in EVD care, clinicians and all the healthcare staff have a critical stake in environmental safety. The routes taken by personnel and patients from low-risk to high-risk areas, water and sanitation, hygiene and waste management are not matters to be left to someone else. A safer environment means safety for patients, healthcare workers, and the community.

The safety of healthcare workers is also, and perhaps principally, related to the proper use of personal protective equipment. Careful and comprehensive training, repeated practice, mentoring by more experienced clinicians, and competency assessment must be in the context of on-site infection prevention and control and clinical procedures, in order to assure safe, sensible, functional, and reproducible practices. A designated controller of the doffing of personal protective equipment and co-supervision in high-risk areas using a buddy system can sharply reduce errors and the risk of contracting infection while caring for patients. Good teamwork among healthcare workers is of vital importance because the safety of each individual depends on the conscientiousness and professionalism of others.

Patient isolation increases the burden of care but also separates patients from their relatives, communities, and healthcare providers. Patient contact with doctors and nurses completely covered by personal protective equipment is very limited, especially because patients typically seek visual contact with doctors/nurses and try to read their facial expressions. This is quite impossible in high-isolation units, rendering the care environment an area where there is scant personal contact. Moreover, the community, which is so important in Africa, can experience the lack of communication with patients, relatives, caregivers, and the world outside the outbreak-affected as traumatic.

EVD is a severe, highly contagious and life-threatening infection. From the beginning of the epidemic, the main objective was to isolate the EVD patients in order to contain the spread of infection. Indeed, several cases were initially due to healthcare transmission, including hospital/outpatient care, home care, and funerary customs. In a retrospective analysis of EVD cases, exposure during traditional funerary rites was cited by 33%, although the proportion of cases reporting this kind of exposure decreased over time, when safe funerary practices and prompt hospitalization contributed to the containment of the epidemic (International Ebola Response Team et al., 2016). However, while bearing in mind that isolation measures/procedures should be strictly followed for suspected/confirmed cases, there is also the need to respect traditions and behaviors, and to communicate safely with healthcare workers, family, and friends. Information, education, and the opening of line-of-sight areas with low barriers where patients can talk with visitors across a safe distance, if possible via electronic communication devices, will generate positive feedback from everyone, relatives/friends and staff. When discussing burial with the families of deceased patients, allow for their viewing of the body and participating in a safe burial. This basic respect for patients and families helps to build and maintain positive relationships with communities, overcoming common misunderstandings and making activities in the EVD care center more transparent (Brett-Major et al., 2015).

FEAR AND STIGMATIZATION OF HEALTHCARE WORKERS WHO COME INTO CONTACT WITH EVD PATIENTS

Management of EVD patients in high-isolation units is challenging and requires the balancing of staff and patient needs. When a healthcare worker is deployed in a resource-constrained and high-risk environment, psychological stress together with physical and emotional fatigue may affect her/his health and contribute to errors that can result in infection. For international staff, the post-deployment period may present additional but under-appreciated stressors. Returning to a higher-resourced healthcare setting leads to the inequity tension experienced by many people working in both economically disadvantaged and affluent countries. Colleagues, neighbors, and others at home may have considerable apprehension about interactions with returning healthcare workers, even though they may have little reason to suspect EVD or other illness. Moreover, whereas healthcare workers are usually well informed about the precautions necessary to prevent the spread of highly contagious diseases, family, neighbors, friends, or non-medical colleagues may erroneously fear that medical personnel might bring infection home from an epidemic overseas. In other words, there is the risk of stigmatization. Fears about Ebola transmission have also caused some government authorities in the United States to enforce the quarantine of volunteer healthcare workers who returned home from West Africa after participating in the treatment of EVD cases, despite CDC recommendations that they should only be actively monitored and not quarantined if they have no fever or symptoms of the disease (McCarthy, 2014).

During the EVD outbreak in Nigeria, there were media reports that patients suspected of having EVD (Ebegbulem, 2014) and healthcare workers who cared for them were feared. Patients suffering from malaria or fever were avoided, abandoned, or rejected by private hospitals in the erroneous belief that they could transmit EVD (Aborisad, 2014). In the Lagos Mainland Hospital, Yaba, where EVD patients were isolated and treated in Nigeria, some healthcare workers who volunteered to treat patients were viewed with suspicion and avoided by colleagues, family members, and the general public. Some were even advised by their families to resign from their roles and return home (Odebode, Adepegba, & Atoyebi, 2014; Ogoina, 2016).

At the beginning of the 2014–2016 Ebola outbreak in Africa, the deployment of international healthcare workers was slow, for three main reasons: lack of information about the situation and how they could help; fear of contracting Ebola; and their families' reactions or resistance to their going (Turtle et al., 2015). Family concern was the main factor that deterred volunteers, followed by the perception of being essential in their current positions (Rexroth et al., 2015).

INTERNATIONAL WORKERS REPATRIATED FROM EVD-AFFECTED COUNTRIES: CLINICAL AND ETHICAL ISSUES

As of June 24, 2015, a total of 65 individuals had been evacuated or repatriated worldwide from the EVD-affected countries. Of these, 38 individuals were evacuated or repatriated to the following European countries: Spain, Germany, France, United Kingdom, Switzerland, the Netherlands, Norway and Italy. The repatriation of international people who voluntarily worked in EVD-affected countries during the epidemic is a response to our social obligation to provide the best available care to aid workers who, in the name of solidarity, are willing to place their lives at risk in order to help others (Kass, 2014). "Whatever you did for one of these least brothers of mine, you did for me" (Matthew 25:40).

Repatriation also benefits the rest of the patients, both by increasing the probability that their caregivers will survive and by decreasing the risk of dissuading other professionals who might otherwise be considering the possibility of traveling to the area in order to help with their care (Donovan, 2014). However, some questions arose when infected workers were repatriated. In

August 2014, a male Spanish missionary nurse was repatriated from Liberia in Spain in order to be treated for EVD, whereas some of his fellow missionaries who belonged to the same religious order and were also infected had to remain in Liberia because they lacked Spanish citizenship (Royo-Bordonada & García López, 2016). The Spanish missionary died a few days later in Madrid, while a female missionary, who had been denied transfer to Spain for treatment, was admitted to the Elwa Public Hospital, a barrack-like building where patients were crowded together without the necessary therapeutic or hygienic conditions, managed to survive the disease, and later walked out of the "death camp," as it was called by the locals (Rego, 2014). The repatriation of volunteers and other workers raises the question of the disparate treatment afforded to international aid workers and to the African health workers on the front lines, who were the principal victims of infection (Rid & Emanuel, 2014).

On the one hand, there is no doubt that the duty of attending to workers who risk their lives falls to their respective countries of origin. Without this "rule," many healthcare workers would be dissuaded from going to resource-constrained countries with an ongoing epidemic. However, the spread of an Ebola outbreak derives in great part from social injustices that do not afford individuals and societies the same level of medical assistance as provided in affluent countries.

Another controversial aspect of repatriation is the risk of introducing an infectious agent into a country that is free of the disease and has little experience in its management. In Spain and the United States, secondary transmissions of EVD to healthcare providers from repatriated patients have been reported, raising the issue of the lack of proper infection prevention and control measures (Ebegbulem, 2015; Parra, Salmeròn, & Velasco, 2014). On the other hand, two Italian healthcare workers with complicated cases of EVD were repatriated to Italy, one of whom received mechanical ventilation in the intensive care unit without any secondary transmission of the virus (Petrosillo et al., 2015).

Finally, yet importantly, the financial cost of repatriation is not negligible. Even if the cost on the ground can be highly variable, the cost of each repatriation is around one million euros (Romero, 2014). For Ebola preparedness during the epidemic, an activity-based cost method was used in the Netherlands, in which the cost of staff time spent in preparedness and response activities was calculated, based on a time-recording system and interviews with key professionals at the healthcare organizations involved. The Dutch healthcare system provided cost information on patient days of hospitalization, laboratory tests, personal protective equipment, as well as the additional cleaning and disinfection required. The estimated total costs averaged 12.6 million euros, ranging from 6.7 to 22.5 million euros. The main cost drivers were personal protective equipment expenditures and preparedness activities by personnel, especially those associated with ambulance services and hospitals. Out of the 13 possible cases that were clinically evaluated, only one confirmed case was admitted to hospital (Suijkerbuijk et al., 2018). The amount of money spent on repatriation appears disproportionate to that given by some European countries for funding prevention projects in EVD-outbreak-affected countries and their border areas (Royo-Bordonada & García López, 2016).

CONCLUSIONS

The recent EVD outbreak in West Africa was unprecedented in scale, larger than all the previous outbreaks combined, and unique in its multi-country spread. Several issues arose from this multi-scientific and multi-cultural experience. Ebola was the embodiment of humanity's fear of what is not fully known and cannot be fought. Our experience has taught us that EVD care should be guided by science, while taking psychosocial and ethical considerations into account, particularly with respect to containing the spread of this highly contagious and life-threatening disease while trying to improve the quality of health care; respecting patients' sensibilities, traditions and customs; minimizing the risks to healthcare workers while providing essential patient care, and determining the optimal solutions for protecting and treating infected local and international healthcare workers who increased their own risk of contracting EVD while saving others' lives.

Health Care Worker (HCW) volunteers from developed countries know that they are exposing themselves to life-threatening risks in order to help patients and professionals in underdeveloped countries. In resource-constrained and high-risk environments they will inevitably experience psychological stress and physical/emotional fatigue. HCWs cannot be immune to fear of contagion in necessarily performing invasive procedures on Ebola patients. But traditional cultures and practices in resource-limited countries exacerbate concerns. Patients may falsely deny previous exposure to EVD and/or traditional funerary rites may lead friends and family to come into contact with the bodies of patients who had died of EVD.

Completely covering HCWs in high-isolation units with personal protective equipment enables them to treat highly contagious EVD patients. But stress and fatigue can contribute to errors and failing to practice optimal safety measures. International HCWs who contracted EVD while voluntarily working in EVD-affected countries were generally repatriated to their countries of origin. It does not mean the stressors were removed. Many died. Family, friends, and associates may be apprehensive about interacting with HCWs returning from treating EDV patients fearing contagion. Given the financial cost of repatriation, the money may have been better used to build facilities in the vicinity of the epidemic, which would have saved many more lives. What is clear is that without international help, the epidemic would have been far worse and more people would have died. Other EVD patients benefitted when caregivers survived and the likelihood that more international HCWs would volunteer to help combat the epidemic was increased. The lessons learned from the world's largest Ebola outbreak must not be lost – these are summarized in Table 1.

Table 1. Ethical Considerations Associated with the Ebola Virus Disease (EVD) Epidemic.

Issues	Consequences
(1) HCWs voluntarily traveled from developed countries to help EVD patients in underdeveloped countries	Without international help, the epidemic would have been far worse and more people would have died
(2) HCWs selflessly exposed themselves to life-threatening risks in order to help others	Many of the HCWs who contracted EVD died
(3) HCWs deployed to resource-constrained and high-risk environments experienced psychological stress and physical/emotional fatigue	Stress and fatigue may have contributed to errors and failure to practice optimal safety measures, resulting in the infection of HCWs
(4) HCWs may discriminate against EVD patients, owing to fear of contagion	Patients may be avoided, abandoned or rejected by HCWs
(5) HCWs performed invasive procedures (intravenous indwelling, hemodialysis, reanimation procedures, mechanical ventilation) on Ebola patients	HCWs were able to save lives but increased their own risk of contracting Ebola
(6) Some patients falsely denied previous exposure to EVD	HCWs and others were placed at risk of infection, some of whom died
(7) Traditional funerary rites required friends and family to come into contact with the bodies of patients who had died of EVD	Many persons were infected due to these practices until safety precautions were instituted
(8) Completely covering HCWs in high-isolation units with personal protective equipment enabled them to treat highly contagious EVD patients	When HCWs' faces were covered with personal protective equipment, their personal contact with patients was diminished
(9) International HCWs who contracted EVD while voluntarily working in EVD-affected countries were generally repatriated to their countries of origin.	• The rest of the EVD patients benefitted because the probability that their caregivers would survive and more international HCWs would volunteer to combat the epidemic was increased • International HCWs infected with EVD received different treatment than the African HCWs on the front lines who were the principal victims of infection • Repatriation of HCWs who contracted EVD introduced the infectious agent into countries that were previously free of it • The financial cost of repatriation was extremely high. The same money could have been used to build facilities in the vicinity of the epidemic, which would have saved many more lives
10. The post-deployment period may present additional but under-appreciated stressors	Returning to a higher-resourced healthcare setting may lead to inequity tension
11. Family members, neighbors, friends or non-medical colleagues may have considerable apprehension about interacting with HCWs returning from treating EDV patients, erroneously fearing contagion	Erroneous fear of contagion has caused returning HCWs to be stigmatized and rejected by family members, neighbors, friends, non-medical colleagues and the general public

Notes: EVD – Ebola virus disease; HCWs – healthcare workers.

REFERENCES

Aborisad, S. (2014). Ebola: FG to meet owners of private hospitals. Punch Newspapers. Retrieved from https://www.africanewshub.com/news/1814659-ebola-fg-to-meet-owners-of-private-hospitals

Brett-Major, D. M., Jacob, S. T., Jacquerioz, F. A., Risi, G. F., Fischer, W. A. 2nd, Kato, Y., … Fletcher, T. E. (2015). Being ready to treat Ebola virus disease patients. *The American Journal of Tropical Medicine and Hygiene*, *92*(2), 233–237.

Centers for Disease Control and Prevention (CDC). (2014). Ebola Virus Disease (EVD). Retrieved from http://www.cdc.gov/vhf/ebola/index.html

Donovan, G. K. (2014). Ebola, epidemics, and ethics—what we have learned. *Philosophy, Ethics, and Humanities in Medicine: PEHM*, *9*(15), 1–4.

Ebegbulem, S. (2014). Anxiety in Edo community as man dies of suspected Ebola symptoms. Vanguard News. Retrieved from https://www.vanguardngr.com/2014/08/anxiety-edo-community-man-dies-suspected-ebola-symptoms/

Ebegbulem, S. (2015). Harsh claims as nurse sues hospital where she contracted Ebola: She alleges insufficient precautions, deceit by the hospital. *Healthcare Risk Management*, *37*, 37–39.

Ebegbulem, S. (2014). Health worker critically ill, as Ebola exposures prompt flurry of medical evacuations. Center for Infectious Diseases Research and Policy (CIDRAP) Website. Retrieved from http://www.cidrap.umn.edu/news-perspective/2015/03/health-worker-critically-ill-ebola-exposures-prompt-flurry-medical

Ebegbulem, S. (2014). Retrieved from https://www.cdc.gov/vhf/ebola/outbreaks/2014-west-africa/index.html

International Ebola Response Team, Agua-Agum, J., Ariyarajah, A., Bawo, L., Bilivogui, P., Blake, I. M., … Yoti, Z. (2016). Exposure Patterns Driving Ebola Transmission in West Africa: A Retrospective Observational Study. *PLoS Medicine*, *13*(11), e1002170.

Kass, N. (2014). Ebola, ethics, and public health: what next? *Annals of Internal Medicine*, *161*(10), 744–745.

Kreuels, B., Wichmann, D., Emmerich, P., Schmidt-Chanasit, J., de Heer, G., Kluge, S., … Schmiedel, S. (2014). A case of severe Ebola virus infection complicated by gram-negative septicemia. *The New England Journal of Medicine*, *371*(25), 2394–2401.

Liddell, A. M., Davey, R. T. Jr, Mehta, A. K., Varkey, J. B., Kraft, C. S., Tseggay, G. K., … Uyeki, T. M. (2015). Characteristics and Clinical Management of a Cluster of 3 Patients With Ebola Virus Disease, Including the First Domestically Acquired Cases in the United States. *Annals of Internal Medicine*, *163*(2), 81–90.

Lyon, G. M., Mehta, A. K., Varkey, J. B., Brantly, K., Plyler, L., McElroy, A. K., … Emory Serious Communicable Diseases Unit. (2014). Clinical Care of Two patients with Ebola Virus Disease in the United States. *The New England Journal of Medicine*, *371*(25), 2402–2409.

McCarthy, M. (2014). CDC rejects mandatory quarantine for travelers arriving from Ebola stricken nations. *British Medical Journal*, *349*, g6499.

Médecins Sans Frontières (MSF). (2014). MSF Reference Books. Retrieved from http://refbooks.msf.org/msf_docs/en/MSFdocMenu_en.htm

Mora-Rillo, M., Arsuaga, M., Ramírez-Olivencia, G., de la Calle, F., Borobia, A. M., Sanchez-Seco, P., … La Paz-Carlos III University Hospital Isolation Unit. (2015). Acute respiratory distress syndrome after convalescent plasma use: treatment of a patient with Ebola virus disease contracted in Madrid. Spain. *Lancet Respir Med*, *3*(7), 554–562.

Odebode, N., Adepegba, A., & Atoyebi, O. (2014). Ebola: Health workers scared, flee hospital. Punch Newspapers. Retrieved from http://theafricachannel.com/ebola-health-workers-scared-flee-hospitals/

Ogoina, D. (2016). Behavioural and emotional responses to the 2014 Ebola outbreak in Nigeria: A narrative review. *International Health*, *8*(1), 5–12.

Parra, J. M., Salmeròn, O. J., & Velasco, M. (2014). The First Case of Ebola Virus Disease Acquired outside Africa. *The New England Journal of Medicine*, *371*, 2439–2440.

Petrosillo, N., Nicastri, E., Lanini, S., Capobianchi, M. R., Di Caro, A., Antonini, M., … INMI EBOV Team. (2015). Ebola virus disease complicated with interstitial pneumonia: a case report. *BMC infectious diseases*, *15*, 432.

Rego, P. (2014). El milagro de la hermana Paciencia. El Mundo. 31 Aug. Retrieved from http://www.elmundo.es/cronica/2014/08/31/540191d6268e3ef7508b456f.html

Rexroth, U., Diercke, M., Peron, E., Winter, C., an der Heiden, M., & Gilsdorf, A. (2015). Ebola response missions: To go or not to go? Cross-sectional study on the motivation of European public health experts, December 2014. *Euro Surveillance*, *20*(12), 21070.

Rid, A., & Emanuel, E. J. (2014). Why should high-income countries help combat Ebola? *The Journal of the American Medical Association*, *312*(13), 1297–1298.

Romero, V. (2014). Un hospital español contra el ébola. El Mundo. 15 Sep . Retrieved from http://www.elmundo.es/solidaridad/2014/09/15/5412dacce2704e932f8b4582.html

Royo-Bordonada, M. Á., & García López, F. J. (2016). Ethical considerations surrounding the response to Ebola: the Spanish experience. *BMC Medical Ethics*, *17*, 49.

Suijkerbuijk, A. W. M., Swaan, C. M., Mangen, M. J., Polder, J. J., Timen, A., & Ruijs, W. L. M. (2018). Ebola in the Netherlands, 2014-2015: costs of preparedness and response. *The European Journal of Health Economics: HEPAC: health Economics in Prevention and Care*, *19*(7), 935–943.

Turtle, L., McGill, F., Bettridge, J., Matata, C., Christley, R., & Solomon, T. (2015). A survey of UK healthcare workers' attitudes on volunteering to help with the Ebola outbreak in West Africa. *PLoS ONE*, *10*(3), e0120013.

Wolf, T., Kann, G., Becker, S., Stephan, C., Brodt, H. R., de Leuw, P., ... Zacharowski, K. (2015). Severe Ebola virus disease with vascular leakage and multiorgan failure: treatment of a patient in intensive care. *Lancet*, *385*(9976), 1428–1435.

World Health Organization (WHO). (2014). Ebola Virus Disease (EVD). Retrieved from http://www.who.int/csr/disease/ebola/en

ETHICS CHALLENGES IN THE DIGITAL ERA: FOCUS ON MEDICAL RESEARCH

Albena Kuyumdzhieva

ABSTRACT

The chapter deliberates on research ethics and the unanticipated side effects that technological developments have brought in the past decades. It looks at data protection and privacy through the prism of ethics and focuses on the need for safeguarding the fundamental rights of the research participants in the new digital era. Acknowledging the benefits of data analytics for boosting scientific process, the chapter reflects on the main principles and specific research derogations, introduced by the EU General Data Protection Regulation. Further on, it discusses some of the most pressing ethics concerns, related to the use, reuse, and misuse of data; the distinction between publicly available and open data; ethics challenges in online recruitment of research participants; and the potential bias and representativeness problems of Big Data research. The chapter underscores that all challenges should be properly addressed at the outset of research design. Highlighting the power asymmetries between Big Data studies and individuals' rights to data protection, human dignity, and respect for private and family life, the chapter argues that anonymization may be reasonable, yet not the ultimate ethics solution. It asserts that while anonymization techniques may protect individual data protection rights, the former may not be sufficient to prevent discrimination and stigmatization of entire groups of populations. Finally, the chapter suggests some approaches for ensuring ethics compliance in the digital era.

Keywords: Big Data; data protection; public data; privacy; human rights; medical research; misuse of data

Ethics and Integrity in Health and Life Sciences Research
Advances in Research Ethics and Integrity, Volume 4, 45–62

ISSN: 2398-6018/doi:10.1108/S2398-601820180000004004

> The digital future of health has only just begun and in many cases outpaces the policy responses.
>
> M8 Alliance Declaration (World Health Summit, 2017)

INTRODUCTION[1]

Twenty years ago, terms such as digital footprints, big data, cloud storage, digital profiling, and invisible computing were only familiar to a few. The World Wide Web was still in its embryonic phase; Google was in "making." Facebook and Twitter did not exist and wearable devices were not in mass production.

Today, millions of people monitor their health status via numerous health apps and activity trackers, seek medical advice in patients' forums and use social media platforms to share concerns or find empathy. Patients communicate with their doctors via WhatsApp. Viber and Facebook messages tweet the latest developments of contagious diseases (e.g., the recent case with Ebola) and rely on a variety of sensor technologies in their everyday life. Bio-electronic medicines, electroceuticals, and web-connected drug devices are not a futuristic vision but rather fast-approaching reality.

The rapid digital revolution has changed the way people live and communicate, thus creating also research development opportunities that were hardly existing in the "off-line" era. Technological advancements have enabled principal investigators to collect and analyze enormous amounts of health data; easily recruit research participants; detect faster, and monitor better, disease activities and outbreaks (Bond et al., 2013). Data science has become an enabling factor for molecular medicine and Big Data analysis to open the doors to treating even the most severe diseases. Health data sharing in the digital era has thus become a necessity, capable of ensuring better quality of life and healthcare provision.

While such benefits are undeniable, they can occasionally overstep some basic fundamental human rights and come at a "price" to ethics. This is an inevitable outcome of the interactions between technology and the social fabric, which bring "environmental, social, and human consequences that go far beyond the immediate purposes of the technical devices and practices themselves" (Kranzberg, 1995). The ethical dimensions of these impacts have confronted the research community with the challenge to predict, recognize, and diminish intangible digital ethics threats with ethics and legal tools that were established to prevent physical harms in the "off-line" era. Thus, the advancement of data science has presented new challenges in the existing ethics research systems and has called for new reflection on the risks that can be borne by research participants in Big Data research (Metcalf & Crawford, 2016; Mittelstadt & Floridi, 2016).

The present chapter deliberates on some of those risks and suggests approaches for ensuring ethics reflection and data protection compliance in the digital era. The chapter evolves as follows: the next section summarizes the key elements of the EU General Data Protection Regulation as it pertains to health research. The third section discusses the emerging ethics concerns, contemplating the nature of privacy and consequent data protection risks, the role of anonymization, and the issues related to bias and representatives of research involving Big Data and/or online-recruited participants. The fourth section deliberates on possible approaches for the way forward. Finally, the last section provides some concluding remarks.

PRIVACY AND DATA PROTECTION SAFEGUARDS: GLIMPSES OF THE GENERAL DATA PROTECTION REGULATION

Big Data health research has been recognized as an opportunity for improving the quality of healthcare delivery and patients' outcomes (Marjanovic, Ghiga, Yang, & Knack, 2017; OECD, 2017). In the long term it can enable customization of health care and ensure that all medical treatments and products are tailored to the individual patient. Such an opportunity however can only become reality if health data are shared on a large scale, thus enabling the advancement of healthcare research. While the benefit of data sharing may be obvious, the willingness of people to provide their health data can be hardly taken for granted. The latest Eurobarometer survey made this very evident, by revealing that only 21% of EU citizens will willingly provide their *anonymized* data to public authorities or public sector companies for medical research purposes, and only 14% of all respondents will do so for private companies conducting medical research. Almost a quarter of the respondents (23%) would not provide their personal health data under any circumstances (TNS Opinion & Social, 2017). While the survey did not specify the reasons behind such data sharing reluctance, it is reasonable to suggest that some of the citizens' reservations are directly linked to privacy and data protection concerns. Such an assumption reflects the results of the Eurobarometer studies on data protection (2015) and E-privacy (2016) indicating that notwithstanding the high priority that majority of EU citizens give to privacy (TNS Political & Social, 2016), most of them do not feel in full control of their data (TNS Opinion & Social, 2015). If this assumption is correct, data sharing practices can be boosted by finding the right equilibrium between chartering new unexplored scientific territories and protecting the fundamental rights of the individuals. The need for such balance is explicitly embedded in the European Convention on Human Rights[2] and the EU Charter of Fundamental Rights,[3] the latter listing the right to freedom of arts and sciences (art.13) along with the rights to human dignity (art.1), respect of private and family life (art.7), data protection (art.8),[4] and non-discrimination (art.21). These rights entail that while scientific activities should be free from constraints, they should abide by the ethics imperatives embedded in the Charter.

The EU General Data Protection Regulation (GDPR) translates these fundamental principles into practical legal terms and procedures, geared to protect and safeguard individuals' privacy and personal data. In doing so, GDPR acknowledges the importance of data processing for scientific purposes and enables science advancement by providing a number of research-specific derogations. The most significant of them refer to purpose and storage limitations, processing of special categories of data, secondary use, and data subject rights.

GDPR places genetic data, biometric data, and data concerning health in the category of special data, which merit particular protection. The latter entails that processing of such data is generally prohibited, with a few notable exceptions, such as scientific research. The processing of special categories of data, however, is subject to the application of appropriate safeguards, aimed at protecting the fundamental rights and interests of the research participants and ensuring proportionality between the aim of the processing and the nature of the processed data. Member States are empowered to choose to maintain or introduce further conditions or limitations with regard to the processing of genetic, biometric, and health data and adopt derogations for some of the data subjects' rights. The latter implies that researchers, wishing to process these special categories of data, should consult the national legal framework of the country where the research activities take place and align their research with the specific national provisions. They should also reflect on the differences in the national legal regimes and the specific requirements that might exist pertaining to the transfer of such special categories of data.

The notion of "broad consent" is another important feature that GDPR introduces to enable the advancement of science. Consent under GDPR should be "freely given, specific, informed and unambiguous indication of the data subject's wishes by which he or she, by a statement or by a clear affirmative action, signifies agreement to the processing of personal data relating to him or her" (art.4.11).[5] The consent should cover all processing activities, carried out with the same purpose, and should be as specific and as informed as possible to prevent possible function creep. These requirements are set as safeguards against unauthorized and unanticipated further use of personal data once the consent has been obtained from the data subject. Within the framework of scientific research however, GDPR provides specific derogation from the consent attributes, introducing the notion of broad consent. This notion is applicable in cases where it is difficult, if not impossible, to envision all purposes of personal data processing at the moment of data collection. In such cases, research participants may give their consent not for a specific project but to a certain area of scientific research. The utilization of broad consent should be however exceptional and allowed only if research participants are given the possibility to opt in or out of certain research and/or parts of research and provided that their rights are safeguarded by adherence to the ethics standards of scientific research (recital 33 GDPR). The Guidelines on consent (Article 29 WP 2018) provide concrete advice in this regard, stressing that when the purpose of the data processing cannot be specified at the time of the collection, the researchers must seek other ways of ensuring data subjects' rights, such as gathering consent for the specific stages of the research that are already known, and acquiring subsequent consent, as the research develops; provision of information on the advancement of the research and its purposes; providing a comprehensive research plan to the research participants; and anonymization, data minimization, and data security measures.

It is also important to note that the notion of broad consent should not be mistaken with the notion of blanket or open consent. The Guidelines on transparency, prepared by the Article 29 Data Protection Working Party[6] (2018: 9), explicitly note that statements such as "We may use your personal data for research purposes" are not compliant with GDPR as they do not provide sufficient information as to the type of the research that is to be carried out. Moreover, the consent, given by the research participant, does not invalidate the researcher's responsibilities with regard to "the fairness, necessity and proportionality, as well as data quality" (Article 29WP 259 rev.01, 2018:3).

Along with the notion of broad consent, GDPR introduces specific derogations from the principle of storage limitation. The latter requires data to be stored for no longer than absolutely necessarily for the purposes for which the data has been processed. Research data may be however exempt from this principle, provided that appropriate technical and organizational measures are set in place. These longer storage periods may enable further research and verification of the research results.

Another important research exemption relates to the obligation of the data controllers to provide information to the data subjects in the cases where the data have not been obtained from the data subjects directly. The researchers may benefit from this derogation in cases where the provision of information proves impossible or will involve disproportionate effort (art.14.5 (b)). In applying this exemption, the number of individuals involved, the age of the data subjects and the safeguards adopted should be taken into consideration.

These special exceptions do not imply however that the privacy and data protection rights of research participants are less safeguarded. GDPR balances the need for research advancement by making these derogations subject to a number of safeguards. Among them are upholding the principle of data minimization at all times and ensuring that appropriate technical and organizational measures for protecting the rights and freedoms of the data subjects are set in place (art.89). In applying these safeguards, researchers should either anonymize or pseudonymize the research data, provided the purposes of the research can be fulfilled in that manner. Privacy data protection impact assessments and the appointment of data protection officers, along with the possibilities for enforcing sanctions for non-compliance, complement the privacy and data protection safety net.

ETHICS CHALLENGES IN THE ERA OF INTERNET AND BIG DATA

By enabling the processing of large volumes of highly diverse biological, clinical, environmental and behavioral data, in a way that no human can, data science has radically changed the way we think and do medical research. Nowadays, sophisticated algorithms help researchers mine, combine, sort and configure data in a speed and fashion that was unimaginable before. Data are reused and repurposed in many different research projects, often making direct contact between the researcher and the research participant unnecessary. The utilization of such techniques, however useful they may be, requires special ethics attention as it creates considerable distance between the researchers and the research participants, transforming the latter from humans, whose rights should be safeguarded, into data entries, amalgamated in large data sets. The latter may easily lead to violations of the established ethics standards for research with humans, as the lack of personal contact, along with the sheer volume of data, can easily create the feeling that personal data in large data sets are "detached" from their owners and therefore are not subject to special protection. Such misconception is extremely dangerous, as it limits the traditional standards of medical ethics to preventing physical and direct physiological harms and disregards the intangible threats that violation of privacy and data protection rights can bring to the life of individuals and entire groups.[7] Such limited understanding of the harm paradigm can be detrimental, not only for the particular research participants, but also for the entire research community: it impacts negatively on the level of citizen trust and willingness to share data for research purposes. This concern is even more pressing in the field of medical research, where the mass processing of personal data can quickly and easily become socially and ethically problematic and place at risk the development of future innovation (Mittelstadt & Floridi, 2016).

Data Use, Reuse, and Misuse

Among the most challenging ethics questions that pertain to Big Data are the ones related to consent and secondary use of data.

Consent by its nature is study-specific, has a concrete purpose, and covers activities that the research participants are informed and aware of. Its main function is to safeguard the privacy and data protection rights of the individuals, by allowing them to exercise their rights to information and withdrawal from the research activities. This notion somehow "clashes" with the very nature of Big Data analytics, which by aggregating, sharing, and repurposing data may reveal unexpected links between data points and deliver results that could have not been predicted. It is therefore difficult, if not impossible at times, to inform the participants on how their data will be used and in what context. Consent, involving Bid Data analytics is therefore highly problematic, as it can hardly be "informed" if the data subjects cannot receive information about the potential uses of their data at the time when the latter is collected or aggregated (Mittelstadt & Floridi, 2016). And while in some specific cases, GDPR may "open" the notion of consent and make it broader for data collected for research purposes, this is not so in the cases when data are gathered in the framework of commercial activities (e.g., health applications, social media platforms). Using such data for research purposes may therefore raise severe ethics concerns related to the validity of the consent, the involuntary participation in research projects and the ability of the individuals to withdraw from the research. Moreover, the latter challenges the general principle that data subjects should have the right to withdraw their consent at any time and in a manner that is as easy as the one for giving consent (art.7.3 GDPR).

A typical example of questionable use of previously collected data is related to the data stemming from social networks or wearable devices. By subscribing to social media services, web forums, open chats or agreeing to the terms and conditions of wearables (to name few examples), millions of people unknowingly agree to give away substantial amount of their personal and often sensitive data for purposes they are mostly unaware of. The ethics issues arising from the use of such data are further aggravated by the fact that in many cases such data are publicly available and can be easily harvested by researchers who have the tools, the access and the knowledge to collect social media data, created in a highly sensitive context (Boyd & Crawford, 2012).

The utilization of such data is highly controversial, as the mere fact of posting information on any media platform does not mean that these data become freely available to everyone. Such an explanation was provided by Danish researchers, who in 2016 published entire data sets, containing usernames and highly sensitive personal data (e.g., drug use, political and sexual preferences) belonging to 70 000 users of the dating site OkCupid. The data were scraped from the profiles of the users for the purposes of conducting psychological study and were published with the justification that the data sets, which were "only systemizing data" from a publicly available website, should be easily accessible for further research purposes.[8]

The OkCupid study represents a particularly worrisome example of how the privacy of individuals can be hijacked and how their basic human rights can be violated under the umbrella of promoting "open science." The right to freedom of science has been interpreted by these particular researchers as an absolute right, overriding the right to privacy and data protection. The obvious lack of informed consent in such cases is well illustrated by Preston, who argues that "people treat social media a bit like they treat the pub…They feel that if they go into a pub and have a private conversation, it does not belong to the pub; it is their conversation. They interpret Twitter or Facebook in the same way – as a place to have a conversation" (House of Commons Science & Technology Committee, 2014:19). The principle that in many jurisdictions people cannot be listened to/filmed or photographed without their consent is therefore fully applicable to the use of online social media data. The misconception of some researchers as to what data are freely available for further use and what data are not, as Mittelstadt and Floridi (2016) argue, derives from profound confusion between the concepts of "being public," referring to making publicly known one's opinion and being "in public," which refers to the concept of being publicly visible. Such misconception is extremely dangerous, as it enables tracking, profiling and analyzing individuals based on publicly available data, without their prior consent and even knowledge. This approach ignores the extreme invasiveness of such research methodology, which has the potential for causing significant harms to individuals and entire communities[9] (see also Metcalf & Crawford, 2016). The misinterpretation of these concepts has been further deepened by the assumption that publicly available data pose only marginal risks to the individuals. Such assumption stems from the binary understanding that data are either public or private, ignoring the fact that in reality, data are dynamic, easily repurposed and in most cases contain personal data, even when they are public. Researchers should be well aware that not all data that are publicly available can be used and shared by everyone (see the definition of open data[10]).

The argument that data from social media platforms are public because the social media users have agreed to the terms and conditions of service and have given their informed consent as to how their data may be used by the third parties should be therefore handled very carefully. As Boyd and Crawford (2012: 672) rightly argue, "[it] may be unreasonable to ask researchers to obtain consent from every person who posts a tweet, but it is problematic for researchers to justify their actions as ethical simply because the data are accessible. Just because content is publicly accessible does not mean that it was meant to be consumed by just anyone." In dealing with the ethical aspects of research involving publicly available data and determining if the data are "open" for use or whether they are private (hence privacy of the individual should be respected), Townsend and Wallace (2016) suggest that the researchers should take into consideration the online environment of the data posting; the reasonable expectations for privacy on behalf of the user (e.g., password protected profiles or closed group discussions) and the ethics obligations to seek consent for the data subjects. Moreover, even if perfectly legal and legitimate, the personal "privacy waver" given by one can potentially influence the rights of others, as the information provided by one participant (via his/her profile) often gives information and access to his/her friends and "friends circle." The latter is violating the privacy of third parties who have not been informed and have not consented to take part in the research activities. The 2018 privacy scandal with the consultancy company Cambridge Analytica, which scraped the personal data of approximately 50 million Facebook users, by allegedly gaining authorized access to the profiles of 270,000 users, provides a grave example how personal data may be misused and the privacy of third parties violated.[11]

These ethics problems are aggravated by the changing functionality of the privacy settings of the different media tools, which may lead to involuntarily sharing of data and potential involvement of unwilling research participants in scientific activities. A typical example of the latter is the transfer of personal customer data from WhatsApp to Facebook. Following the 2014 acquisition of WhatsApp by Facebook, the message platform changed its terms of service, in a way that enabled the transferring of the personal data of its clients to Facebook. Such transfer affected enormous amount of individuals, who may have deliberately chosen to stay outside of Facebook for privacy or other concerns.[12]

The debate on how to ethically use data collected via social media tools is broadened further by the lack of clear consensus regarding whether the deletion of a post or account equates to withdrawal from research and the discussion related to the right of parents to provide their children's data and the actual ability of the children (once matured) to withdraw their data or avoid possible negative consequences (e.g., child health data that can lead to potential discrimination/stigmatization of the child in his/her adulthood). In dealing with such challenges, the researchers should apply the principle that "the insights and the requirements identified by ethical inquiry precede the rules of law" (Goodman, 2015: 121).

To the above ethics dimensions, we should also add a new one, which is often neglected by the researchers: the media where the sensitive personal information is stored or processed, or the software provider that is used for data analysis. Such concern is particularly pressing in the light of the IBM X-Force Research (2017) on security threats in the healthcare industry, revealing that one of the biggest healthcare data breaches for the past five years was a security breach in a software service provider company, which led to exposure of healthcare data of almost four million individuals.

Ethics concerns as to the adequacy of the data and privacy protection may arise also for service providers which are located outside of the EU and not covered by adequacy decisions.[13] Researchers should be aware that cloud service or software providers based outside of the EU do not always guarantee adequate protection of personal data, compliant with GDPR. The use of such services means that the personal data of EU citizens may be processed without further control and can be even sold if appropriate derogations from the commercial terms and conditions are not set in place (Bruin & Floridi, 2017). Researchers should therefore give careful considerations and ensure that their data processing contracts with third parties provide adequate GDPR compliant protection or, if this is not the case, ensure that the research participants are fully aware of the privacy/security risks to their data.

Online Recruitment

Social media has been arguably among the most effective online tools for the recruitment of participants for youth-related studies (Amon et al., 2014; Leonard et al., 2014). Yet, despite its effectiveness, major questions arise as to determination of the actual age of the participants and the difference that may exist between their actual "off-line" age and their "on-line" self-determined age. Back in 2011, a consumer report[14] study revealed that 7.5 million Facebook users were under the age of 13, and more than 5 million of them were 10 and under. These numbers are only likely to have increased in the past years. This phenomenon is highly problematic from a research ethics perspective, as, on the one hand, it may lead to the research participants' exposure to potential physiological or privacy risks, stemming from research methods that are not appropriate for their age. On the other hand, the reliability of the research results might be jeopardized by the absence of valid consent.

Researchers may also face challenges in studies with the participation of elderly people or certain types of vulnerable groups, who may be able to consent but may not be entirely used to the new technologies and may not have full understanding how their personal data may be used, reused or misused. In such cases, it is of paramount importance to give appropriate consideration to the individual characteristics of the particular research group and its cultural context, even in cases when the research activities do not infringe particular law or terms of use.

Bias and Representativeness of Research

Research, involving Big Data and/or online recruitment of participants, faces ethics challenges related not only to the privacy and data protection but also to the representativeness and potential bias of the research results. Social media platforms and other IT tools provide access to unprecedented amounts of data and potential research participants, thus facilitating research advancement. Yet, these resources should be treated with special care, as drawing conclusions based solely on online sources of participation can jeopardize the research conclusions and lead to reduced representativeness, self-selection bias, sampling bias, and other reliability issues (Alshaikh, Ramzan, & Rawaf, 2014; Khazaal et al., 2014). The research risks derive from three major characteristics of the online environment: the first one is that some population groups may be underrepresented (such as those with poor literacy) and others may be more online active and particularly interested in taking part in certain research activities (Denecke et al., 2015; Phillips, 2011). The second one is related to possible differences between the online and off-line social behavior, and the third one relates to the possibility for one user to have multiple accounts or one account to be used by multiple users, a fact that may question the overall reliability of the research (Boyd & Crawford, 2012).

Data taken out of its context raises additional concerns, as it may significantly impair the objectivity of the research findings. The latter, along with the sheer volume of the analyzed data, may often lead to "seeing patterns where none actually exist, simply because enormous quantities of data can offer connections that radiate in all directions" (Boyd & Crawford, 2012: 668). Moreover, if the appropriate methodology is not set in place, the risk of questionable outcomes is higher for studies that involve Big Data analysis, as the risks of having "big noises" and thus big errors are quite substantial. Algorithmic grouping and profiling in the context of medical research may easily lead to potential stigmatization, discrimination or even harm to entire groups of the population, based on the real or perceived interconnections, such as, for example, religious or ethnic origin and violence (Mittelstadt & Floridi, 2016). To prevent such outcomes, Boyd and Crawford (2012) advocate that researchers working with Big Data should understand not only the limits for interpreting data but also the limits of the data sets and the questions they can ask from them.

DEALING WITH RESEARCH ETHICS IN THE DIGITAL ERA

The utilization of novel data science methods and practices has come to challenge well-established medical ethics traditions by questioning the applicability of the classical definitions of "human subjects" and "interventions" and highlighting the power asymmetries between large-scale data sets studies and individuals' rights to data protection, human dignity and respect for private and family life. In many cases, those power asymmetries, along with the invasiveness of some Big Data research methodologies, raise severe privacy and data protection concerns. At the time when the "power follows the control of data" and data have become "the new oil" (High Level Group of Scientific Advisors, SAM, 2017: 23), substantial attention should be given to the ethics implications that Big Data research may have (Metcalf & Crawford, 2016; Mittelstadt & Floridi, 2016).

Is Anonymized Data the Ultimate Solution?

The use of anonymized data for research purposes is a default principle embedded in art.89 of the GDPR. The latter requires the use of anonymized data, if the purposes of the research can be fulfilled in that way. Processing of personal data in research context should therefore only take place if the purposes of the research cannot be reached by processing anonymized or pseudonymized data.

At first glance, anonymizing data in an attempt to avoid any ethics and data protection compliance responsibility seems like the best solution. The processing of anonymous data[15] is not covered by the GDPR framework and is not usually subject to any ethics review. Moreover, as data mining is hardly perceived as an "intervention" in the life or body of the individual, it thus falls outside of the scope of the traditional medical ethics review process.

Seeing anonymized data as the ultimate solution, which may take the "ethics burden" off the shoulders of the researchers, may prove to be somewhat problematic for many reasons. Some of them refer to the technical aspects of the anonymization process and to drawing a clear distinction between pseudonymized and anonymized data. The lack of proper understanding of the technical aspects behind these concepts may easily mislead some researchers that they do not deal with personal data, thus exposing the privacy and data protection rights of the individuals at risk.

The fact that data are anonymized may solve some of the data protection requirements but may not be the ultimate remedy for ethics. The latter is clearly identified by Phillips, Borry and Shabani (2017) in their systematic literature review on the need for institutional ethics approvals for research involving anonymized data and samples. The study identifies privacy and re-identification risks; inadequacy of anonymization procedures; the inability to prevent possible group ethics harms and the higher risks associated with genomic research among the main ethics risks pertaining to the use of anonymized data for research purposes. These concerns are also clearly spelled out by the USA National Institute of Health, which notes that "it may be possible to re-identify de-identified genomic data, even if access to data is controlled and data security standards are met, confidentiality cannot be guaranteed, and re-identified data could potentially be used to discriminate against or stigmatize participants, their families, or groups. In addition, there may be unknown risks" (National Institutes of Health. 2015:2). In the same vein, the Irish Council for Bioethics (2005) acknowledges that research involving anonymous materials can protect the anonymity of the particular individual, but not the one of the group to which he or she belongs. Given the possible risks for group discrimination or stigmatization, the Irish

Council of Bioethics advises that research involving the use of anonymous archival biological material should be subject to institutional ethics review.

Big Data analytics only aggravate the re-identification concerns, as the merger of otherwise anonymized data sets can easily lead to re-identification through cross referencing. Such a possibility jeopardizes the ability of the individuals to exercise their right to privacy and control over their personal data. A typical example is the anonymous data collected by wearables, which, when combined with location data or/and other identifiers, can be traced back to particular individuals and reveal their geographical location, daily habits, activities and health data.[16]

Ethics Approach toward Data: Some Suggestions

Kranzberg's first law reads: "Technology is neither good nor bad; nor is it neutral" (Kranzberg, 1995). It creates a new context within which the values and the ethics principles of the society should be embedded and the potential impact on humans should be assessed. The first step in dealing with the new ethics challenges in medical research is therefore the acknowledgment that the use of new technologies does not change the existing ethics principles, standards, and regulations. The fact that researchers may not have a face-to-face interaction with the research participants, but deal with their health data instead, does not make the rights of the individuals less valid or less protected. On the contrary, this remote connection makes individuals more vulnerable, as in many cases they are not in full control of their data and may not be in a position to enforce their right to privacy, data protection, and non-discrimination.

To address this potential vulnerability, the General Data Protection Regulation places special emphasis on the responsibility of the data controllers and processors to safeguard the rights to privacy and data protection. Translated in practical terms, GDPR requires researchers and research institutions to embed the principles of privacy and data protection in the design of their studies and ensure that privacy and data protection by default are applied for all research participants. This approach, as Chassang (2017) argues, "approximates the law and the technology, two essential elements of the data protection system that shall develop together to allow legal compliance in a modern world." The software and hardware industry and IT research enterprises will therefore play a key role in assisting the research community, by developing and designing services, applications, and hardware solutions that meet the data protection requirements "by design" and "by default." The latter will further promote adherence to the highest standards of ethics, thus providing human rights protection that is not always available in all parts of the world.

What is more, the protection of personal data has been now linked to a system of sanctions, which provides sufficient compliance incentives even for the largest and richest data controllers and processors. GDPR provides mechanisms for enforcing stronger compliance, by envisioning administrative fines for infringements of the data protection rights of individuals going as high as 20,000,000 EUR or 4 percent of the global annual turnover (whichever is higher). GDPR introduces also the obligations for all public authorities and for entities, which in the execution of their core activities process on large-scale special categories of personal data (e.g., hospitals), to appoint an institutional data protection officer (art.37 (1). It also requires research institutions to carry out data protection impact assessment when the data processing may bring high risk for the rights and freedoms of the data subject[17] (for all conditions see art.35 (1). While such requirements may be seen by some as burdensome, the conduct of data protection impact assessment and the appointment of an institutional data protection officer are of key importance for the advancement of the research and knowledge economy and reinforcement of trust and integrity of science. From an ethics perspective, the evaluation of the ethics risks related to the data processing operation can bring significant benefit for the researchers, as such evaluation will enable them to map properly their personal data processing activities, reflect on the ethics risks and ensure the highest standards of ethics compliance and research participants' protection.

Reflecting on the need for providing effective protection to the rights of privacy, data protection and non-discrimination, Denecke et al. (2015:145) suggest that any research study "should pre-plan in its design and pre-empt any ethically problematic effects introduced by its users, application area, and relevant dimension, and should develop sound solutions for addressing these ethical issues." In the same vein, Harvard Catalyst Regulatory Foundations (2015) suggests three simple steps for accommodating the peculiarities of social media research: (1) adherence to the existing rules and regulations; (2) assurance that the proposed media recruitment techniques comply with the policies and the terms of use of the relevant websites and affirmation that the proposed recruitment strategy is sensitive to the privacy of the potential participants; (3) respect the norms of the community and uphold its trust in science. This also implies that, when dealing with personal data, researchers should adhere to the principle of contextual identity and only collect data that are absolutely needed for the given research and use them exclusively in that context (unless explicitly permitted by the data subject). Such an approach "should prevent personal data which is divulged with permission in one context from being used in a totally different one, without the individual's consent or knowledge; for example, personal medical data being used for commercial purposes" (High Level Group of Scientific Advisors, SAM, 2017:43).

Researchers should also be aware of the potential invasiveness of use and misuse of publicly available data and the ethics harms that utilization of such data may bring. Central place in the ethics discussion should be given not only to questions as "what kind of data is obtained and how" but also to questions as "what is done with this data and what type of insights may be generated" (Metcalf & Crawford, 2016).

Another key feature for preventing ethics harms for research, involving anonymized data, is the assessment of the technology and the security measures set in place to guarantee that the research data are truly anonymous. Such measures are among the CoE recommendations on the research on biological materials of human origin (CoE, 2016), which call for verification of the non-identifiability of the used biological samples and prescribes that "[n]on-identifiable biological materials may be used in a research project provided that such use does not violate any restrictions defined by the person concerned before the materials have been rendered non-identifiable and subject to authorisation provided for by law" (art.21(4). Elliot et al.'s (2016) publication on anonymization techniques and anonymization decision-making framework can be (among others) a useful starting point in addressing these process requirements.

The utilization of Big Data for research purposes poses also significant challenges for Research Ethics Committees, which historically have been set to prevent physical harm to individuals and safeguard the welfare of the research participants. In the digital era, they have to deal with much less tangible threats and predict potential ethics harms stemming from the unpredictable results of Big Data research. To do so, Research Ethics Committees may be faced with the need to identify unethical practices embedded in various algorithms or identify data mining, merging, and anonymization techniques that may pose ethics threats and reveal the identity of the individuals in otherwise anonymized data sets. Along with the traditional knowledge in ethics and medical studies, these specific tasks require profound understanding of the new technologies, their operational capacities and data processing procedures. It thus asks for inclusion of data scientists and/or IT ethicists in the Committee's composition. Such an approach will also facilitate discussions about the potential invasiveness of use and misuse of publicly available data and raise the awareness about the ethics harms that the utilization of such data can bring. The role of the Research Ethics Committees thus should go beyond the traditional ethics review and extend to translating bioethics into data science and providing support in modeling practices which sustain public trust and encourage health data sharing across countries and disciplines.

The development of codes of conduct and guidelines tailored to the specific research areas can also boost ethics compliance and address some of the dilemmas faced by the researchers. A good step in this direction is the BBMRI-ERIC[18] initiative for establishing a Code of Conduct for Health Research. The aim of the Code is to provide sector-specific advice and explain in easily accessible format the practical steps for ensuring compliance with GDPR, taking into account the peculiarities of processing data for purposes of scientific research in the area of health.

The Recommendations from the Association of Internet Researchers Ethics Working Committee (2012) on the ethical decision-making and internet research represent another good tool developed to support researchers using internet-based data sources. Rather than prescribing a number of ready-made solutions, this document provides a list of questions that should be addressed by internet researchers and those who supervise them.

At the level of system design, the Institute of Electrical and Electronics Engineers (IEEE) contributes to reconciling ethics and technology by developing a Model Process for Addressing Ethical Concerns during System Design. The aim of the standard is to provide engineers and technologists with design approaches and software engineering methods that take into consideration and minimize ethics risks from the very onset. The standard will also enable easier identification of potential ethics threats and facilitate the utilization of technological solutions which embed ethics by default and by design.

CONCLUSIONS

Data science is enabling us to process data in a way that was unimaginable 20 years ago. It has helped science to advance in many research fields and has ultimately changed the ways research is conducted. In doing so, it has also challenged the traditional ways ethics is applied.

Handling ethics issues in the digital era may, at first sight, look challenging and overly complicated. Yet, the answers to the ethics issues that occur today have the same objectives as in the times past: protection of the basic fundamental human rights and minimizing any potential risk or harm (current or/and future) for the research participants and their families. Therefore, responding to ethics issues should not be regarded as an unnecessary burden, but rather as an indispensable and key component for enhancing the value of science and the esteem with which scientists are held.

"Ethics, like writing computer code or rendering a nursing or medical diagnosis, is sometimes easy and sometimes not. In either case, however, one might get it wrong" (Goodman, 2015:14). To get it right, particular attention should be paid to the principle of proportionality, the right to privacy, the right to the protection of personal data, the right to the physical and mental integrity of a person, the right to non-discrimination and the need to ensure high levels of human health protection (art.19 Horizon 2020 Regulation), regardless of the "online" or the "off-line" nature of every research. The use of publicly available data should be carefully reviewed, from an ethics perspective, throughout its use.

With the lack of generally accepted standards, capable of prescribing the desired ethical approach for each particular situation, the ethics compliance of each research project should be measured by the extent to which it respects the dignity and rights of every individual and adheres not only to the letter of the law but also to its spirit.

NOTES

1. The content of this article does not reflect the official opinion of the European Union. Responsibility for the information and views expressed lies entirely with the author.
2. Council of Europe (1950) Convention for the Protection of Human Rights and Fundamental Freedoms.
3. Charter of Fundamental Rights of the European Union (2000/C 364/01).
4. The right to data protection is also included in the Treaty on the Functioning of the European Union (Article 16.1).
5. For research involving clinical trials, the processing of data should also comply with the requirements established in the Regulation (EU) No 536/2014 of the European Parliament and of the Council of 16 April 2014 on clinical trials on medicinal products for human use, and repealing Directive 2001/20/EC.
6. Article 29 Data Protection Working Party an advisory body, comprised by representative from the data protection authority of each EU Member State, the European

Data Protection Supervisor and the European Commission. Following the entry into force of GDPR, the body is transformed into the European Data Protection Board.

7. For example, research linking intelligence, health and ethnicity may lead easily to discrimination or stigmatization of the individuals or even the entire group of people

8. Retrieved from http://fortune.com/2016/05/18/okcupid-data-research/

9. Boyd (2007) provides examples of possible group ethics harms, by deliberating on research linking health data with census data on ethnicity. He argues that while the outcomes of such study may bring potential health benefits for ethic minority groups (e.g. if unusually high incidence of, or predisposition to certain diseases is discovered), depending on the research context, it may also lead to possible stigmatisation of the group, if appropriate ethics safeguards are not installed.

10. According to the definition of Open Data Institute, "Open data and content can be freely used, modified, and shared by anyone for any purpose." For more information see http://opendefinition.org/

11. For more information: https://www.nytimes.com/2018/03/19/technology/facebook-data-sharing.html. Accessed on March 27, 2018.

12. For more information: https://www.cnil.fr/en/data-transfer-whatsapp-facebook-cnil-publicly-serves-formal-notice-lack-legal-basis. Accessed on March 27, 2018.

13. European Commission has the power to determine whether a third country ensures an adequate level of protection. If a third country is covered by adequacy decision, personal data may flow freely between EU and the third country, without any further safeguard being necessary.

14. Retrieved from https://www.consumerreports.org/media-room/press-releases/2011/05/cr-survey-75-million-facebook-users-are-under-the-age-of-13-violating-the-sites-terms-/

15. Anonymised data is data that does not relate to an identified or identifiable person or is rendered anonymous in such a manner that the data subject is not or no longer identifiable (rec.26, GDPR). Council of Europe (2016) in its recommendation on research on biological materials of human origin comes to a similar definition, stipulating that "'non-identifiable biological materials' are those biological materials which, alone or in combination with data, do not allow, with reasonable efforts, the identification of the persons from whom the materials have been removed" (art.3.1ii).

16. See fitbit data use: https://www.cbsnews.com/news/pentagon-reviews-fitness-tracker-use-over-security-concerns-fitbit/. Accessed March 27, 2018)

17. Particular examples of cases where impact assessment is required involve processing of large scale of sensitive data or activities related to systematic monitoring of publicly accessible areas on a large scale.

18. Biobanking and BioMolecular Resources Research Infrastructure – European Research Infrastructure Consortium.

REFERENCES

Alshaikh, F., Ramzan, F., & Rawaf, S. (2014). Social Network Sites as a Mode to Collect Health Data: A Systematic Review. *Journal of Medical Internet Research*, *16*(7), e171.

Amon, K., Campbell, A., Hawke, C., & Steinbeck, K. (2014). Facebook as a Recruitment Tool for Adolescent Health Research: A Systematic Review. *Acad Pediatric*, *14*(5), 439–447.

Article 29 Data Protection Working Party. (2018). Guidelines on Consent under Regulation 2016/679, WP259, rev.01. Retrieved from http://ec.europa.eu/newsroom/article29/item-detail.cfm?item_id=623051

Article 29 Data Protection Working Party. (2018). Guidelines on transparency under Regulation 2016/679, WP260.rev.01. Retrieved from http://ec.europa.eu/newsroom/article29/item-detail.cfm?item_id=622227

Association of Internet Researchers. (2012). Ethical Decision-Making and Internet Research: Recommendations from the AoIR Ethics Working Committee (Version 2.0). Retrieved from https://aoir.org/reports/ethics2.pdf.

Bond, C., Ahmed, O., Hind, M., Thomas, B., & Hewitt-Taylor, J. (2013). The conceptual and practical ethical dilemmas of using health discussion board posts as research data. *Journal of Medical Internet Research, 15*(6), e112.

Boyd, D., & Crawford, K. (2012). Critical Questions for Big Data, Information. *Communication and Society, 15*(5), 662–679.

Boyd, K. (2007). Ethnicity and the ethics of data linkage. *BMC Public Health, 7*(318).

Bruin, B., & Floridi, L. (2017). The Ethics of Cloud Computing. *Sci Eng Ethics, 23*(1), 21–39.

Chassang, G. (2017). The impact of the EU general data protection regulation on scientific research. *Ecancermedicalscience, 11*(709), 1–12.

Council of Europe. (1950). Convention for the Protection of Human Rights and Fundamental Freedoms as amended by Protocols No. 11 and No. 14.

Council of Europe. (2016). Recommendation of the Committee of Ministers to member States on research on biological materials of human origin, CM/REC (2016) 6. Retrieved from https://search.coe.int/cm/Pages/result_details.aspx?ObjectId=090000168064e8ff

Denecke, K., Bamidis, P., Bond, C., Gabbarron, E., Househ, M., Lau, A., Mayer, M., Merolli, M., & Hansen, M. (2015). Ethical Issues of Social Media Usage in Healthcare. *Yearbook of Medical Informatics, 10*(1), 137–147. Retrieved from https://www.ncbi.nlm.nih.gov/pmc/articles/PMC4587037/

Elliot, M., Mackey, E., O'Hara, K., & Tudor, C. (2016). *The Anonymisation decision-making framework*. UK: UNCAN Publications, Retrieved from http://ukanon.net/ukan-resources/ukan-decision-making-framework/

Goodman, K. (2015). *Ethics, Medicine, and Information Technology*. Cambridge: Cambridge University Press.

Harvard Catalyst Regulatory Foundations. (2015). *The Use of Social Media in Recruitment to Research: A Guide for Investigators and IRBs*, Retrieved from https://catalyst.harvard.edu/pdf/regulatory/Social_Media_Guidance.pdf. Accessed on April 4, 2017.

High Level Group of Scientific Advisors, Scientific Advice Mechanism. (2017). *Cybersecurity in the European Digital Single Market*. Brussels: DG RTD, European Commission.

House of Commons Science and Technology Committee. (2014). *Responsible Use of data, Fourth Report of Session 2014–15*. London: The Stationery Office Limited.

IBM X-Force Research. (2017). *Security trends in the healthcare industry. Data theft and ransomware plague healthcare organizations*. Retrieved from https://public.dhe.ibm.com/common/ssi/ecm/se/en/sel03123usen/SEL03123USEN.PDF.

Irish Council for Bioethics. (2005). *Human biological material: recommendations for collection, use and storage in research*. Retrieved from http://health.gov.ie/wp-content/uploads/2014/07/Human_Biological_Material1.pdf

Khazaal, Y., Singer, M., Chatton, A., Achab, A., m Zullino, D., Rothen, S., Khan, R., Billieux, J., & Thorens, G. (2014). Does self-selection affect samples' representativeness in online surveys? An investigation in online video game research. *Journal of Medical Internet Research, 16*(7), e164.

Kranzberg. (1995). Technology and History: "Kranzberg's Laws. *Bulletin of Science, Technology & Society, 15*(1), 5–13.

Leonard, A., Hutchesson, M., Patterson, A., Chalmers, K., & Collins, C. (2014). Recruitment and retention of young women into nutrition research studies: practical considerations. *Trials, 15*(23), 1–7.

Marjanovic, S., Ghiga, I., Yang, M., & Knack, A. (2017). *Understanding value in health data ecosystems. A review of current evidence and ways forward*. UK: RAND Corporation.

Metcalf, J., & Crawford, K. (2016). Where are human subjects in Big Data research? The emerging ethics divide. *Big Data & Society*, January–June 1–14.

Mittelstadt, B., & Floridi, L. (2016). The Ethics of Big Data: Current and Foreseeable Issues in Biomedical Context. *Science and Engineering Ethics, 22*(2), 303–341.

National Institutes of Health. (2015). *Guidance on consent for the future research use and broad sharing of human genomic and phenotypic data subject to the NIH genomic data sharing policy*. Retrieved from https://osp.od.nih.gov/wp-content/uploads/NIH_guidance_elements_consent_under_gds_policy.pdf

OECD. (2017). Ministerial Statement. The Next Generation of Health Reforms, OECD Health Ministerial Meeting, 17 January 2017. Retrieved from https://www.oecd.org/health/ministerial-statement-2017.pdf

Phillips, A., Borry, P., & Shabani, M. (2017). Research ethics review for the use of anonymized samples and data: A systematic review of normative documents. *Accountability in Research*, *24*(8), 483–496.

Phillips, M. (2011). "Using social media in your research. Experts explore the practicalities of observing human behavior through Facebook and Twitter." *American Phycological Association*. Retrieved from http://www.apa.org/gradpsych/2011/11/social-media.aspx. Accessed on April 4, 2017.

Regulation (EU) 1291/2013 of the European Parliament and the Council. (2013). Establishing Horizon 2020 – The Framework Programme for Research and Innovation (2014–2020) and repealing Decision No 1982/2006/EC.

Regulation (EU) (2016). 2016/679 of the European Parliament and of the Council of 27 April 2016 on the protection of natural persons with regard to the processing of personal data and on the free movement of such data, and repealing Directive 95/46/EC, Official Journal of the European Union, 119, 4.5.2016. General Data Protection Regulation.

TNS Opinion & Social. (2015). Data Protection Report, Special Eurobarometer 431.

TNS Opinion & Social. (2017). Report Attitudes towards the impact of digitisation and automation on daily life, Special Eurobarometer 460 Wave EB87.1. Retrieved from https://ec.europa.eu/jrc/communities/sites/jrccties/files/ebs_460_en.pdf

TNS Political & Social. (2016). Report E-Privacy, Flash Eurobarometer 443. Retrieved from http://ec.europa.eu/COMMFrontOffice/publicopinion/index.cfm/Survey/getSurveyDetail/instruments/FLASH/surveyKy/2124

Townsend, L., & Wallace, C. (2016). Social Media Research: A Guide to Ethics, The University of Aberdeen. Retrieved from https://www.gla.ac.uk/media/media_487729_en.pdf

Wave EB83.01. Retrieved from http://ec.europa.eu/commfrontoffice/publicopinion/archives/ebs/ebs_431_en.pdf

BIG DATA IN HEALTHCARE AND THE LIFE SCIENCES

Janet Mifsud and Cristina Gavrilovici

ABSTRACT

Big Data analysis is one of the key challenges to the provision of health care to emerge in the last few years. This challenge has been spearheaded by the huge interest in the "4Ps" of health care (predictive, preventive, personalized, and participatory). Big Data offers striking development opportunities in health care and life sciences. Healthcare research is already using Big Data to analyze the spatial distribution of diseases such as diabetes mellitus at detailed geographic levels. Big Data is also being used to assess location-specific risk factors based on data of health insurance claims. Other studies in systems medicine utilize bioinformatics approaches to human biology which necessitate Big Data statistical analysis and medical informatics tools. Big Data is also being used to develop electronic algorithms to forecast clinical events in real time, with the intent to improve patient outcomes and thus reduce costs.

Yet, this Big Data era also poses critically difficult ethical challenges, since it is breaking down the traditional divisions between what belongs to public and private domains in health care and health research. Big Data in health care raises complex ethical concerns due to use of huge datasets obtained from different sources for varying reasons. The clinical translation of this Big Data is thus resulting in key ethical and epistemological challenges for those who use these data to generate new knowledge and the clinicians who eventually apply it to improve patient care.

Underlying this challenge is the fact that patient consent often cannot be collected for the use of individuals' personal data which then forms part of this Big Data. There is also the added dichotomy of healthcare providers which use such Big Data in attempts to reduce healthcare costs, and the negative

Ethics and Integrity in Health and Life Sciences Research
Advances in Research Ethics and Integrity, Volume 4, 63–83

ISSN: 2398-6018/doi:10.1108/S2398-601820180000004005

impact this may have on the individual with respect to privacy issues and potential discrimination.

Big Data thus challenges societal norms of privacy and consent. Many questions are being raised on how these huge masses of data can be managed into valuable information and meaningful knowledge, while still maintaining ethical norms. Maintaining ethical integrity may lack behind in such a fast-changing sphere of knowledge. There is also an urgent need for international cooperation and standards when considering the ethical implications of the use of Big Data-intensive information.

This chapter will consider some of the main ethical aspects of this fast-developing field in the provision of health care, health research, and public health. It will use examples to concretize the discussion, such as the ethical aspects of the applications of Big Data obtained from clinical trials, and the use of Big Data obtained from the increasing popularity of health mobile apps and social media sites.

Keywords: Big Data; health care; health research; ethics; life sciences; genomics

WHAT IS BIG DATA AND WHAT DOES IT MEAN FOR HEALTH PROFESSIONALS AND RESEARCHERS?

Our society has become "data rich and data dependent" (Rodriguez, 2013). The demand and interest in collecting and analyzing more and more data from more people has led to a type of data fetishization (Lupton, 2014). Individual cases, even with rare diseases, are less attractive, even the scientific journals are reluctant in publishing case reports, and favor the publication of massive data results. We live in the age of "intensified data sourcing," a concept that has emerged as a new way of running health services. This has been defined as attempts at getting more data, of better quality, about more people (Hoeyer, 2016). Traditional approaches to health research generally rely on a small quantity of data, collected in highly controlled circumstances, such as randomized clinical trials. With Big Data, techniques such as the extraction of large amounts of data facilitate inductive reasoning and an exploratory analysis of data (Roski, Bo-Linn, & Andrews, 2014).

Big Data analysis is one of the key challenges to the provision of health care and life science to emerge in the last few years. For health professionals, Big Data generated by patients becomes a new source of information, likely to enrich diagnostics and patient monitoring. Biomedical Big Data can be expressed in many forms such as genetic sequencing data, electronic health records aggregated clinical trials, and biological specimens. In this digital era of medicine, which tends to become increasingly personalized, the data may be collected directly from healthcare providers, but also from health applications, wearable tracking technologies, point-of-care diagnostics, monitoring technologies, research institutes, epidemiological centers, pharmaceutical laboratories, imaging centers, hospital reports, insurance companies, as well as social networks and forums.

Big data can be stored in biobanks, or virtual research repositories, taking the form of aggregated datasets.

The necessity of providing Big Data for research purposes was born in the essence and strategy of the "precision medicine," a concept that has thoroughly promoted the idea of an "omics-driven" medicine (Ginsburg, 2014). This challenge has been spearheaded by the huge interest in the 4Ps of health care (predictive, preventive, personalized, and participatory).

Healthcare research is already using Big Data to analyze the spatial distribution of diseases such as diabetes mellitus at detailed geographic levels. Big Data is also being used to assess location-specific risk factors based on data of health insurance claims or to develop electronic algorithms to forecast clinical events in real time, with the intent to improve patient outcomes and thus reduce costs. Large datasets can also be used to uncover the effects of exposures that may have small consequences on individuals but large cumulative effects on populations or to identify specific subgroups (unlike the clinical studies where it may be difficult to recruit enough participants to be able to identify significant differences) or to study rare conditions. Furthermore, administrative records can be used to track individuals over time, and so are ideally suited to measuring the long-term impacts of health conditions or interventions (Currie, 2013).

The capacity to link phenotypic and clinical data with whole-sequenced genomes is the first step in the transition to a model of health care based on prediction, prevention, and personalization (Hockings, 2016). The use of genomic information is at the forefront of the Big Data debate, as it is increasingly recognized that DNA-based information or single-nucleotide polymorphism (SNP) profiles can potentially be used to identify individuals. Big Data provides researchers the ability to link records from a number of data sources, thus offering great promise for a better understanding of disease or disease predictors. It also may harm individuals by uncovering negative or potentially discriminatory health-related findings (Salerno, Knoppers, Lee, Hlaing, & Goodman, 2017). Other studies in systems medicine utilize bioinformatics approaches to human biology which necessitate "Big Data" statistical analysis and medical informatics tools.

Large datasets are already being used to support translational research such as the UK's Genomics England, China's Kadoorie Biobank, and the US Precision Medicine Initiative. The pharmaceutical industry has also set up major collaborations to tackle the challenge such as AstraZeneca with the Wellcome Trust and the Sanger Institute with San Diego-based biotech Human Longevity (Morrison, Dickenson, & Lee, 2016). New infrastructures are also being developed to facilitate the sharing of scientific data such as the transnational Biobanking and Biomolecular Resources Research Infrastructure (BBMRI-ERIC), the Human Heredity and Health in Africa (H3Africa) network, and the Global Alliance for Genomics and Health (GA4GH) for genomic and clinical data (Morrison et al., 2016).

Yet, this Big Data era also poses critically difficult ethical challenges, since it is breaking down the traditional divisions between what belongs to public and to private domains in health care and health research. Increasing ethical concerns are being raised: transparence, accountability, respect for privacy, etc. If two decades ago we were talking about autonomy, privacy, justice and equity, today "new" values come into play: reciprocity, solidarity, citizenry, or universality (Knoppers & Chadwick, 2005). These challenges are being widely discussed in specialized circles which focus on the emerging challenges of biobanking and Big Data through the linkage of biospecimens and medical and nonmedical data (Morrison et al., 2016).

There is a huge bottom-up pressure for Big Data to be used which is being driven by both patients and healthcare systems. It is also key to identify primary, secondary, and tertiary sources used in forming these Big Datasets in order to understand what could be the potential pitfalls, external constraints, and social influencers which will influence the extrapolation of this Big Data to actual healthcare applications (Balas, Vernon, Magrabi, Gordon, & Sexton, 2015). For example, electronic health records in large datasets have been shown to include as much as 4.3–86% of incomplete and inaccurate data in various fields. Even the patients' activism associations – this form of "social movement" – have become important partners in the biomedical research community, with a key role in the mobilization of patients as research participants, but also in raising capital for research itself (Rabeharisoa et al., 2014). In their discourse, Big Data is valued within the concept of data sharing.

Underlying this challenge is the fact that patient consent often cannot be collected for the use of individual data which then forms part of this Big Data. There is also the added dichotomy of healthcare providers which use Big Data in attempts to reduce healthcare costs, and the negative impact this may have on the individual with respect to privacy issues and potential discrimination.

Big Data thus challenges societal norms of privacy and consent. Many questions are being raised on how these huge masses of data can be managed into valuable information and meaningful knowledge, while still maintaining ethical norms. Maintaining ethical integrity may lag behind in such a fast-changing sphere of knowledge. There is also an urgent need for international cooperation and standards when considering the ethical implications of the use of Big Data-intensive information.

DATA OWNERSHIP, SHARING, AND STORAGE

The collection and analysis of massive data from electronic health records created the opportunity to build the information infrastructures that can facilitate the sharing of personal health data (Gherardi, Østerlund, & Kensing, 2014), and thus producing a mutual advantage. If a platform for data sharing is perceived to only promote the benefits of its owners (usually the researchers or big companies), it will lose its social value. However, the benefit provided to platform participants does not have to consist of money or some other individual reward, but in providing participants with an opportunity to act altruistically and to feel the worthiness of their actions (Riso et al., 2017). This is why data sharing has been compared with other actions that promote the general public good and social rewards such as blood donation.

Thus, data repositories have been perceived as a "common good" resource. The value attached derives from the capacity to deliver information related to human life and disease susceptibility; therefore, they are globally relevant. The population biobanks store genetic and non-genetic information (e.g., medical history, demographics). The next-generation sequencing technologies like genome-wide association studies (GWAS) use this information through analysis of statistical patterns of genetic variation to find genetic correlates for common diseases.

Yet, data storage and the consequent processing of personal data raise several concerns, primarily related to the balance between individual rights and interests and societal benefits. More practically, this is about balancing privacy and data protection against other societal values such as public health, environmental protection, security, or economy efficiency (Riso et al., 2017).

CORE VALUES ATTACHED TO THE USE OF BIG DATA

It has been argued that data sharing is not a value-neutral practice, but one shaped by a broad spectrum of ethical, political, and social goals and the electronic platforms reflect diverse conceptions of data sharing among stakeholders (Riso et al., 2017). Big Data imposes new reflections upon our values, how we carry out our actions, and how it gives a larger number of people more means to communicate and interact with each other. One may ask whether these "massive data" raise specific ethical issues. In this section we will identify and analyze the core values essential for the ethical sharing of Big Data.

Scientific Value

Scientific value in data sharing platforms refers to the need for quality, quantity, and accessibility of data. Data sharing has a moral worth when adequate standards are fulfilled. Data quality concerns accuracy, reliability, completeness, consistency (OECD, 2013; Rippen & Risk, 2000), and awareness of the introduction of bias in the dataset by means of the ways the data are collected (Leonelli, 2014). Data quantity relates to the ability to collect and connect large datasets. Obviously, a greater dataset may lead to better interpretations than a smaller one. However, the collection of large quantities of data may end up being less accurate, less reliable, incomplete, and of less consistent data quality (Mayer-Schönberger & Cukier, 2014). Data accessibility relates to the ability of different health organizations or individuals to exchange information accurately and to use this information (Heubusch, 2006). One should keep in mind that the greater the opportunities for sharing data, then the greater will be the opportunities for health data to be misused. Therefore, the scientific value is related to the benefit of the research using Big Data. Objective criteria like the amount and statistical power of a data collection, the quality, and the level of data elaboration in research might be helpful for assessing the potential benefit of data.

Risk of Harm

If the reporting of research conducted using routinely collected health data is poor, or inappropriate methods were used to obtain them, then important findings may be lost. As a result, resources may be allocated to suboptimal care (Vayena, Salathe, Madoff, & Brownstein, 2015). The risk of harm in research on Big Data is strongly related to data protection. For decades, the use of clinical records in research has been agreed on the basis that removing common identifiers such as names and Social Security numbers from the data *nearly* eliminates the risk of harm. This approach, once called "anonymization," is currently known as "de-identification" (reflecting thus the probabilistic nature of re-identification). When data are de-identified, anonymity is not, technically speaking, guaranteed: instead, identifiers are removed or masked to the point where the probability of re-identification appears very small (Kulynych & Greely, 2017).

Respect for Person and Informed Consent (IC)

Surveys show that people are willing to share their genetic data and biospecimens, provided that their permission to do so is sought (Wellcome Trust, 2013). By enabling greater access to personal data concerning one's health and genetic information, the risk that personal health data could accidentally be disclosed or distributed to unauthorized parties emerges (Hoffman, 2010). There is broad agreement that individuals should control their own data and make decisions about accessing the data, and how the data will be used (Sterkckz, Rakic, & Cockbain, 2016). This also implies the possibility of withdrawing the information, which sometimes could not be guaranteed since the process of sharing data tend to mix the different kinds of data through different platforms and to upload it in complex networks (Shabani & Borry, 2015).

Autonomy means the capacity of individuals to be self-governing, to make decisions, and to act in accordance with them and with their values, commitments, and life goals (Global Network, 2015). Blasimme and Vayena (2016) explored the implications of precision medicine initiatives with respect to current understandings of autonomy. They reiterated that greater participation in decision-making may also result in more meaningful choices to participants involved in medical research.

In order to promote autonomy, the platforms should envisage clear and full meanings of communication, education, and informed consent. Autonomy will be violated if the human samples or data are being used without consent, even if the purpose is to help others (e.g., to help people suffering from rare diseases). Privacy is related to the value of autonomy. Even if privacy is invaded with no detrimental consequences, the very fact that people have not been asked permission infringes upon their autonomy. This may have negative repercussions on the levels of trust in the healthcare system.

Informed consent, as the "agreement to a certain course of action, such as treatment or participation in research, on the basis of complete and relevant information by a competent individual without coercion" is intended to prevent the unauthorized usage of a participant's own data (Global Network, 2015).

In the framework of Big Data debate, autonomy is not always considered the most important value, although the bioethics today tends to place it in the core of any healthcare decision (Dawson, 2010). Instead, a public health-oriented ethics, where the population health prevails, might offer a new perspective in the age of Big Data. It might be argued, for instance, that when the physical and privacy risks for participants are negligible, proper safeguards are in place, and data are being used to promote public health (or other public goods), the secondary use of data for purposes other than those initially intended is morally acceptable (Global Network, 2015). Thus, the balance between individual autonomy and public good needs to be carefully assessed.

In the context of Big Data research, informed consent raises the following challenge: while there is an unequivocal obligation to obtain consent, trying to obtain explicit written or witnessed verbal consent from bio-bank participants, which might be in the hundreds of thousands, would not be feasible. This is particularly difficult, since the information gathered by biobanks is intended for use in a range of future projects that are not known at the time individuals consented to participation. Furthermore, it is well known that an important requirement of informed consent is that subjects have the right to withdraw from participation at any point during the study. However, in the case of research on anonymized Big Data, anonymization prevents any influence the donors might claim on the use of their samples and makes it practically impossible to re-contact participants for obtaining consent or returning clinically significant findings (Goodman, 2016; Rothstein, 2010).

No consensus seems to exist on a theoretical level whether "blanket consents," "specific consents," "no consents," or "broad consents" fit best as a consent model for balancing the interest of donors and research in the best possible way. Some support the rights of research participants to remain autonomous and in control of their data; others encourage the societal benefit and propose to keep data usage in the hands of the scientists and ethics committees, as long as participants agree. In practice, however, the model termed "broad consent" has been adapted by many current biobank projects (Master, Nelson, Murdoch, & Caulfield, 2012). Broad consent is not an open or a blanket consent. To give a broad consent means consenting to a framework for future research of certain types. Included in this framework is ethical review of each specific research project by an independent ethics committee as well as strategies to update regularly the biobank donor and ongoing withdrawal opportunities. If anything in the framework changes, the participant should re-consent. In that sense, broad consents still claim to be informed consent (Helgesson, 2012). In the end, the potential societal benefits of access to data will compensate ethical concerns about informed, autonomous decision- making.

Broad consent has been challenged. Its ethical acceptability has been a source of serious debates. Some question even if it really constitutes an informed consent. While initially it was seen as a necessary compromise for addressing the consent of biobanks, it has been rejected by some scholars, and considered as an unacceptable ethical solution, claiming that it does not in fact satisfy the high ethical standard set by fully informed consent. It was argued that one cannot possibly consent to future research projects, often unspecified and to some extent unforeseen, therefore hindering donors from exercising fundamental rights and freedoms (their autonomy). Furthermore, the promoters of this view render "informed broad consent" to Big Data research a contradiction in terms (Karlsen, Solbakk, & Holm, 2011; Steinsbekk, Kåre Myskja, & Solberg, 2013).

Consequently, an approach termed "dynamic consent" has been proposed to meet such challenges. This is a web-based platform with an interface that allows research participants to have an "interactive relationship with the custodians of biobanks and the research community" (Kaye, 2012). The dynamic process is accomplished by a continuous re-contact with biobank donors, giving them "real-time" information on specific research projects and enabling the participants to easily provide or revoke their consent (Whitley, Kanellopoulou, & Kaye, 2012). In contrast to a broad consent model (where the patient has a passive implication), a dynamic consent involves narrower, more specific consents with active opt-in requirements for each downstream research project, thus being a bidirectional, ongoing, interactive process between research participants and researchers. The ethical attention in this way remains focused on what the individual decides, not what the individual has at stake. A dynamic consent would alleviate participants' fears by putting them in control of their data. This new form of consent has not remained unchallenged. Steinsbekk et al. (2013) consider that it is more ethical to restrict the choices of those who agree to participate in longitudinal biobank research than to present them with a whole range of options that a dynamic consent could offer. In other words, presenting participants with the information necessary to allow them to make informed decisions about how their samples and data are used would make them to "raise their guard" instead of having them trustful of research ethics committees and researchers. The perceived problem with dynamic consent is that few participants will probably be able to fulfill the expectations placed upon them to make informed decisions. Thus, having more options will ultimately lessen the availability of data.

There are six major advantages with dynamic consent:

(1) It fulfills better the specifications of autonomy embedded in informed consent, by giving information for new types of research in real time rather than asking for broad consent at the beginning of the research (Kanelloupoulou et al., 2011).
(2) It keeps participants better informed by continuously sending out updated information about specific research projects via SMS, e-mails, or websites (Steinsbekk et al., 2013).
(3) Increased engagement in biomedical research (particular for people who in general are skeptical and would otherwise not contribute to biobank research) (Kaye, 2012).

(4) Increased participant control by engaging public in biobank governance (through "proposing, drafting or amending governance structures, protocols, strategies, and policies") (Dove, Joly, & Knoppers, 2012).
(5) It transfers ethical responsibility from committees to participants (since new consents will be asked for each new project, thus eliminating the need for review boards or ethics committees to reassess the validity of a previously given consent) (Steinsbekk et al., 2013).
(6) It enables the return of research results and incidental findings to participants.

However, there might be several weaknesses with a dynamic consent approach as well: a dynamic strategy for consent process might inherently link research with health care, especially when it promises the return of individual research results, carrying thus the risk of therapeutic misconception around the purpose of research participation (Steinsbekk et al., 2013). Furthermore, there is always the risk of participants not opting in or the risk of disappointment due to unfulfilled expectations.

However, some commentators point out that not allowing broad consent would inhibit the willingness of many participants to share their data (Dove et al., 2012).

Finally, the Institute of Medicine (2015) draws a clear distinction between "interventional research" and research that is "exclusively information based." It recommends that the latter type of research be allowed without individual consent as long as researchers have policies and procedures in place to protect data privacy, and as long as the research is done for clearly defined and approved beneficial purposes.

Trust

The opaque nature of the data collection process undermined the trust in the organization acting as custodians of the data, and the lack of information regarding the further use of the data would further erode this trust. Trust has been perceived as the end-point of the obligation to share research data with other scientists (Brakewood & Poldrack, 2013). Therefore, today trust is an essential and indispensable value for sharing of health data. It is tightly connected to transparency, the accessibility of information, and accountability – the existence of clear procedures related to health data sharing as well as the capacity of an institution to provide a compensation or protection to parties, when harmed (by accident or negligence) (Prainsack & Buyx, 2013).

Trust is particularly relevant in Big Data research using routinely collected health data where individual consent to participate is not usually possible (Beauchamp, 1996), and where the open nature of data collection means that information on the purposes and analysis of collected data are stipulated in a broad sense (ter Meulen, Newson, Kennedy, & Schofield, 2011).

Appropriate and transparent reporting may allow individuals (including lay people) to assess the quality of research on biomedical Big Data, to understand how empirical evidence contributes to scientific decision-making and policy drafting, making the research process opened up to scrutiny. This will increase the public awareness on the research purposes, on how data are being used for research, as well as on the results of the research (Vayena et al., 2015). Increased transparency of research, assuring visibility or accessibility of information, is possible through complete and accurate publication of research (Davenport et al., 2010). As Elliott and Resnik (2014) have argued: "Society is likely to be better served when scientists strive to be as transparent as possible" (Elliott & Resnik, 2014).

Benefit

Benefit is considered as one of the most important criteria for valuable research, being the basic justification for the duty to share data. The potential benefits of Big Data sharing are difficult to predict or to measure. The main benefit of data sharing derives from the potential to produce valid and reliable scientific results and to maximize the uptake of findings that will improve health outcomes, thus producing a potential therapeutic benefit.

Objective criteria like the amount and statistical power of a data collection, the quality and the level of elaboration of the data, or the rarity of the data in the research community might be helpful for assessing the potential benefit of data. The greater the benefit, the stronger is the data producers' duty to share (as owned to the public and to the data donors).

The benefit value in data sharing also refers to efficiency and distributive justice. To ensure a fair balance of interests, proper mechanisms for benefit sharing should be put in place when designing data sharing platforms (Chadwick & Berg, 2001). Benefit is a particularly important value when considering the involvement of vulnerable populations in research. They may be put at risk by being included in Big Data repositories (e.g., due to privacy loss) while they may not be able to benefit from the knowledge generated. For example, the disadvantaged members may not benefit from the new drugs and services due to their high price which may not be covered by universal healthcare services and insurance systems (Global Network, 2015). This is an issue of justice in relation to benefit sharing to a vulnerable population.

Sustainability

This appears to be one of the most difficult values to achieve, being close to the principles of equity and efficiency. Sustainability in the context of Big Data sharing refers to the expectation that the platform will continue to deliver its services after a research project ends, thus justifying also the costs of building it. In order to grant sustainability, one needs to provide the long-term resources required as well as information related to termination or transfer of ownership or control of the platform, if applicable. Sustainability requires the ability of the resource to update the content and to accommodate novel needs (Vassilakopoulou, Skorve, & Aanestad, 2016).

However, a platform that fulfills the criteria of transparency, user empowerment and benefit sharing, might not be feasible, in the sense of sustainability, because it is not perceived as an attractive investment by the public or private entities that could fund it (Riso et al., 2017). An electronic platform may also depend on (unusual) cultural, political, or financial circumstances, which should not be ignored and might be perceived as a weakness. This is why a sustainable platform should not rely exclusively on public funds or private investors but mostly on the revenues it generates from the wider benefits to the entire research and industry ecosystem (Harris et al., 2012).

Equity, Justice, and Fairness

Equity refers to the distribution of rights and obligations related to contributions, maintenance, and use of a common good. Who does the work and who earns the benefits? Is the resource readily available when needed for all members of the community? Are all voices heard when significant decisions are made regarding the governance of the common good?

With respect to fairness, researchers benefit from the work of others from which they draw ideas, methods, or data and on which they can build. As such, the principle of fairness (strongly linked with reciprocity) requires them to also contribute to this pool of knowledge so that others may benefit (Carter, Laurie, & Dixon-Woods, 2015). Fairness also applies in the context of research publication, as the obligation of researchers not to be biased in the presentation of findings or selection of evidence.

Distributive justice in research on Big Data is linked to the transparency of the research report. Failure to adequately report may impede on reproduction or replication of results, particularly if this leads to an unjustified burden on some members of society. For example, some groups may be inappropriately identified as being affected by a disease – and thus stigmatized – when a failure to report transparently hinders the replication of results and produces an inefficient use of resources (Vayena et al., 2015).

Solidarity

Solidarity has been thoroughly acknowledged when discussing the biobanks governance, particularly in relation to rare diseases. It has been defined as "people's willingness to carry costs (financial, social, emotional, or otherwise) to assist others." While solidarity overlaps with the concept of charity, reciprocity, altruism, or empathy, it has not been perceived as a moral value or as a political concept, but more as a set of practices that apply at three levels:

(1) interpersonal level (born from the fact that people face similar risks, share the same experience, or work toward the same goal);
(2) community level – solidarity as shared practices in a given group or community (these are more institutionalized solidarity-based practices, informal groups and arrangements that are not (yet) consolidated by contractual or legal arrangements); and

(3) institutional level but having a contractual relationship (Prainsack & Buyx, 2013).

Reciprocity

When talking about data sharing, some of the most significant values are reciprocity and societal incentives. Riso et al. (2017) considers that society could facilitate reciprocity by providing infrastructures (platforms) with proper safeguards, making it easy to access and share data while protecting both scientists and donors. This could be achievable by requiring platforms to demonstrate how this contributes to the common good – for instance, by allowing anonymized data to be exported to public institutions for public health purposes. Also, by requiring platforms to provide a high degree of autonomous control of the provided personal, society will promote reciprocity.

It has been debated that secondary data users might appeal to a duty of solidarity and reciprocal support among researchers. However, it is questionable whether the secondary data users have a genuine right to access data. Data producers own the sharing of data to the public, in order to benefit the public and advance science. The secondary data users have a rather "instrumental" role with respect to benefiting the public and sciences.

It has been argued that secondary data users' appeal to reciprocity is also objectionable for there is no guarantee of real reciprocity in international data sharing practices among researchers (Sane & Edelsteinm, 2015). Furthermore, one may be aware, for instance, that requiring certain levels of reciprocity between data sharers may be a moral requirement, but at the same time it might inadvertently exclude researchers from resource-poor countries (Boddington, 2012). Reciprocity goes along with the notion of data sharing benefits, where the benefits produced by the data should be shared among all parties involved in the production of that data (platform users and owners).

Honesty

The honesty of the individual is related to personal obligations not to deliberately deceive readers. Furthermore, this may also refer to "intellectual honesty," meaning to be straightforward with the description of the research hypothesis, analysis, interpretation, and reporting of results (Masic, 2012). In the context of genomic sequencing one should notice the concept of "radical honesty," according to which individuals volunteer de-identified genetic information for public sharing (Hayden, 2012). Another implication of honesty in research on Big Data is related to distributive justice. For instance, funding a study that duplicates an existing, but poorly reported, study means that other research is not funded. Such unfunded research could yield potentially useful results and as such there may be benefits that are lost by not undertaking this research.

Big Data also requires new understandings of concepts such as *robustness, resilience, adaptability,* and *usability.* Individuals need the assurance that when asked to consent to sharing data in the short term, this reconciles with ongoing robustness and resilience together with regulatory oversight for long-term use.

CLAIMS AGAINST DATA SHARING

Costs and Burdens for Data Producers

While the costs and burdens associated with all working steps necessarily involved in data sharing should be minimized, according to the data producers' research resources, one may argue the greater the costs and burdens for data producers, the more it is ethically challenging to *require* data sharing.

The Risks for Data Donors

When sharing the data with secondary data users on a global scale, one should consider the potential harms from breaching confidentiality and abuse of their data. The greater the presumed risks for data donors, the less weight to data donors' intention to promote science has to be granted and the less the public has a right to require data sharing. Therefore, before making ethically problematic use of data, one needs to be sure whether there are feasible and less problematic alternatives (Schickhardt, Hosleym, & Winklerm, 2016).

BIG DATA IN HEALTH CARE: TRANSLATION IN HEALTH CARE

The focus on 4Ps in health care is set to radically change the way health care is provided, disease is treated, and medicine is practiced. These 4Ps represent:

- Predictive: Identification of individual risks of developing certain diseases based on the person's genetic profile and other personal information.
- Preventive: Methods and treatments to avoid, reduce, and monitor the risk of developing certain diseases.
- Personalized: Clinical interventions based on the unique genetic, medical, and environmental characteristics of each patient-citizen, and genomic profile of his/her diseases.
- Participatory: Citizens are fully engaged in personal health management.

These 4Ps are challenging societal norms of privacy and consent. Many questions are being raised about how these 4Ps can be addressed through the transformation of huge masses of data into meaningful knowledge while still maintaining ethical norms. There is an urgent need for international coordination to consider the ethical implications of data-intensive information in order to maintain research integrity. Big Pharma have shown great interest in Big Data. Already in conditions such as epilepsy, an industry-ICT collaboration explored the application of applied machine learning to large claims databases to construct an algorithm for antiepileptic drug (AED) choice for individual patients (Devinsky et al., 2016a). Based on this, a model-predicted AED regimen with the lowest likelihood of treatment change was assigned to around

10,000 patients following a training group of 40,000 patients and outcomes were evaluated to test the validity of the model. It was found that treatment was more successful if patients received the model-predicted treatment. Thus, this Big Data model could enable personalized, evidence-based epilepsy care, and thus better health outcomes for patients.

Fischer, Brothers, Erdmann, and Langanke (2016) posed epistemological questions that "systems medicine" and the use of bioinformatics tools and algorithms raise for patient care. They reiterate the importance of addressing possible unforeseen consequences which may arise due to the clinical ambiguity in genomic data-sharing and digital data collection from wearable smart devices. Ethical safeguarding of this data in retrospective research needs to be developed. Safeguards should also be put in place to ensure that there is no risk of exacerbating existing inequalities or ability to redress any grievance, especially in developing countries. Complicating the picture further is the fact that expectations for health care varies considerably across disease groups and the envisioned roles of end users (patients, clinicians, and those mediating between groups). While this highlights the need for inclusive, deliberative models of governance that incorporate the needs of different stakeholders in the translational process, studies have shown that there are some common areas of concern, such as the likely cost of new personalized treatments and how this new information might affect access to health care.

BIG DATA IN PUBLIC HEALTH

There are three main stakeholder types: individuals who wish to keep track of data about their health; researchers who use these data; and commercial entities who collate the data to develop novel self-tracking devices, apps, or services (Bietz, et al., 2016). Key questions about the dichotomy between data privacy and data which is in the public interest is challenging traditional understandings of transparency and consent. Epistemological understandings of societal norms and values such as individual versus institutional power, identification versus identity, and virtual versus real individuals are being redrawn.

The controversial use of health data in the UK lead to a huge social media outburst on Twitter. The Twitter comments were analyzed around key themes which gave an indication of public perception on the subject (Hays & Daker-White, 2015). An online survey carried out among 465 individual and 134 researchers in the US (Bietz et al., 2016) sought to identify key challenging issues for the use of this Big Data. These were found to include data ownership; data access for research; privacy; informed consent and ethics; research methods and data quality; and digital applications with devices – apps were highlighted. The respondents stated that they would be willing to share their data if it was used to advance research for the good of the public. Researchers were keen to use Big Data for their research, but acknowledged that challenges existed with respect to intellectual property, licensing, and the need for clear legal basis and that too many controls can create barriers to Big Data research. Commercial entities reiterated the importance of having good customer relationships (Balas et al., 2015).

New computing safeguards are needed to address public concerns such as that research access would only be to aggregated data and not to individually identifiable information. However, it is also recognized that limitations may arise. For instance, in the Devinsky Antiepileptic Drug Study (Devinsky, Hesdorffer, Thurman, Lhatoo, & Richerson, 2016b) it was recognized that population demographics differed from those normally seen in epilepsy clinics. This may have biased the algorithm's prediction of AEDs. Moreover, the model could not evaluate seizure freedom, reductions in seizure frequency, or quality of life. The limitations were partly offset by the power of the large patient population.

BIG DATA IN HEALTH RESEARCH

Individualized findings generated through Big Data approaches can be applied directly in clinical decision-making. Yet it is important to ensure that this information is translated appropriately. The use of Big Data–driven scoring systems may raise ethical issues. One example is the Riyadh Intensive Care Program Mortality Prediction Algorithm, (sometimes referred to as the "death computer"). The use of these predictive clinical algorithms may challenge the epistemological basis of a doctor–patient relationship. The use of scoring data from incidental and secondary findings such as from genomics, imaging and patient recorded data may also have implications of derived from the joint analysis of data from diverse sources (Fisher et al., 2016). It is important that this data is accurate and valid and that public opinion is continuously considered in order to prevent a lack of trust in the healthcare system (Hays & Daker-White, 2015).

BIG DATA IN GENETIC DATA

While all that we have discussed so far applies for any types of data, tensions emerge in relation to genetic data ownership and sharing. The openness toward data sharing in genetic research became evident in 1996, when according to Bermuda principles, established within the Human Genome Project (1996), the DNA sequence data should be published and/or uploaded to public repositories within 24 hours of production. Thus, data could be shared even before publication, not only post-publication as was the common practice (Contreras, 2011). In the research field, funders and publishers of research proposals are in a position in which they can enforce data sharing. Curators have been hired to check that each paper's variant descriptions have been accurately transmitted to a public database.

Genetic data governance arrangements impact the way genetic testing results could be used for a much wider population than the one affected by rare inheritable diseases. The need to set up strict data governance is becoming stringent since the volume of genetic data is growing exponentially as genetic tests for hereditary conditions are becoming less costly, private laboratories are expanding their offers of diagnostic and prognostic tests, and direct-to-consumer testing is becoming a growing business. The next-generation sequencing techniques provide a massive amount of genetic data, useful for both clinical and research purposes. Usually these are combined with other types of data, like health records, physical location, or socio-economic status extracted from different registries (Hood, Lovejoy, & Price, 2015). Most often stored separately on research servers, the genomic data is likely to remain linked to patient identities and their medical data and preserved for future needs as researchers find new, disease-linked variations in the human genome (Lunshof, Chadwick, Vorhaus, & Church, 2008).

Aside from potential clinical uses, gene sequencing is commonly used in research, often without the specific consent of the persons whose DNA is sequenced. In fact, the scientists who have access to previously collected samples of a patient's blood or tissue often sequence the patient's genome without asking consent to sequencing. At best, the patient whose clinical specimens are sequenced for research may have signed a clinical consent form, often containing a general disclosure statement according to which the samples and data may be shared for unspecified future research. Informed consent for genetic biobanking can never be adequate enough, since it is given by individuals, while the samples reveal information about families and groups (Beskow et al., 2001). Autonomy will be violated if the human samples or data are being used without consent, even if the purpose is to help others.

Furthermore, this will be an intrusion into people's privacy. The large-scale genomic research, based on the provision of vast amounts of genetic information, was perceived as a threat to genetic privacy. The privacy-laden character of genetic information derives from the fact that genetic information was compared with the person's "probabilistic future diary." But even if privacy is invaded with no detrimental consequences, the very fact that people have not been asked permission infringes upon their autonomy. This again may have negative repercussions on the level of trust in the healthcare system. Protection of genetic privacy represents a way of minimizing genetic discrimination, meaning to avoid treating individuals less favorably based upon the information derived from their DNA. In this case genetic privacy has instrumental (or extrinsic) value – as it lessens or prevents genetic discrimination.[1] While it is obvious that genetic discrimination is the consequence of breaching the privacy, any unauthorized access to genetic information violates genetic privacy regardless of whether genetic discrimination takes place. From this perspective, the various laws and regulations do not practically limit the research access to genetic information (e.g., the 2008 US GINA – Genetic Information Non Discrimination Act), but they prevent genetic discrimination.

There is a vigorous debate over different aspects of genomic privacy: whether genomic data are identifiable, and whether patients have a moral obligation to participate in such research regardless of personal preference. For a long time, it was considered that the genome contains "de-identified" information, the disclosure of which poses no significant privacy risk. However, nowadays it is debatable whether subjects of genomic research can reasonably expect to remain anonymous, as some new studies suggest future re-identification is increasingly possible (Kulynych & Greely, 2017). Keeping individuals truly de-identified is becoming increasingly difficult with genomic analyses and data gathering.

In this frame, genetic exceptionalism refers to a special protection that genetic data deserves due to its exceptional character, different from other kinds of health-related information (Suter, 2004). The roots of genetic exceptionalism trace back in 1984 when the President's Commission on Bioethics Report concluded that genetic information should not be released without first obtaining "explicit and informed consent" (Commission for the Study of Ethical Problems in Medicine & Biomedical and Behavioral Research, 1984). During the early 1990s, advances in genetic research motivated the introduction of numerous regulations worldwide aimed at banning potential forms of genetic discrimination. Advocacy groups (such as Working Group of the National Action Plan on Breast Cancer (NAPBC)) worked from the beginning to promulgate the view that genetic information is distinct from any other type of medical information and private, since it provided information about an individual's predisposition to future disease (Hudson, 1995). UNESCO's International Declaration on Human Genetic Data (2003) (IDHGD) reinforced the "special status" of genetic data, due to its sensitive nature related to predictability of genetic predispositions, having thus the capacity to impact on relatives and offspring, extending over generations.

Genetic exceptionalism also means that genetic information has an intrinsic value regardless of consequences; therefore, handling this information requires certain safeguards: more stringent rules for accessing genetic information as compared to medical information generally. However, stricter regulations may hinder, or obstruct, the progress of large-scale genetic investigations but the right to genetic privacy trumps the value of genomic research. The broad use of these large datasets may not match the original purpose for which consent was obtained, thus raising concern for protecting the privacy of research participants (Allain & Ormond, 2015).

CONCLUSIONS

Big Data in health care raises complex ethical concerns due to use of huge datasets obtained from different sources for varying reasons. The clinical translation of Big Data results in key ethical and epistemological challenges for those who use these data to generate new knowledge and who eventually apply it to improve patient care. The main associated ethical issues relate to questions of privacy, autonomy, convenience, surveillance, encryption and anonymization, and ownership. Big Data challenges our understanding of traditional ethical concepts, such as consent, and the right to withdraw, since many large-scale data gathering operations rely on a broad consent to use and reuse data for multiple purposes. While it may be asserted that the associated moral values are already supported by legal frameworks, consistency in their practical implementation may vary considerably. Big Data platforms challenge traditional ideas on how the translational patient pathway should be organized. There is also a perceived danger that could lead to new normative requirements, forcing individuals into thinking that it is a moral duty to participate in such studies, even at the risk of giving away personal information for a greater good.

If public health is a common good from which everyone benefits, then in the current debate over data sharing one should not forget to address how health-related Big Data will be used for the common good, while still respecting individual rights and interests. One should keep in mind that the greater the opportunities for sharing data will be, the greater the opportunities for health data to be misused will also occur.

There are several measures to consider for risk mitigation when sharing the data: the trustworthiness of the secondary data user and their institution, the accountability and data protection standards of the secondary data user's institution, and the data protection law to which secondary data users and their institutions are subject.

In the field of Big Data related to genomic research, the risk related to breach of privacy is strongly connected with potential discrimination; therefore, we need to grant the exceptional character of genetic data. According to genetic exceptionalism, merely accessing an individual's genetic information without obtaining consent is seen as essentially wrong, regardless of any possible negative consequences. Informed consent, considered the core principle of bioethics, does not appear to seem feasible for data repositories and Big Data methodologies, in its traditional meaning. If broad consent will gain more legal and ethical acceptability, then proof of well-functioning and robust ethical oversight is needed to retain public trust. Transparent and complete reporting of the study results, the methods and data sources, will be imperative for such an assessment.

NOTE

1. Genetic discrimination is different from discrimination based on traits (eg: race) that have a genetic component.

REFERENCES

Allain, D. C., & Ormond, K. E. (2015). Ethical issues in genetic and genomic research. In J. L. Berliner (Eds.), *Ethical Dilemmas in Genetics and Genetic Counseling* (pp. 186–208), Oxford: Oxford University Press.

Balas, E. A., Vernon, M., Magrabi, F., Gordon, L. T., & Sexton, J. (2015). Big Data clinical research: Validity, ethics, and regulation. *Studies in Health Technology Information*, *216*, 448–452.

Beauchamp, T. L. (1996). Moral foundations. In S. Coughlin & T. L. Beauchamp (Eds.), *Ethics and epidemiology* (pp. 24–52). New York, NY: Oxford University Press.

Beskow, L. M., Burke, W., Merz, J. F., Barr, P. A., Terry, S., Penchaszadeh, V. B., … Khoury, M. J. (2001). Informed consent for population-based research involving genetics. *Journal of the American Medical Association*, *286*(18), 2315–2321.

Bietz, M. J., Bloss, C. S., Calvert, S., Godino, J. G., Gregory, J., Claffey, M. P., Sheehan, J., & Patrick, K. (2016). Opportunities and challenges in the use of personal health data for health research. *Journal American Medical Information Assocation*, *23*(e1), 42–48.

Blasimme, A., & Vayena, E., (2016). Becoming partners, retaining autonomy: Ethical considerations on the development of precision medicine. *BMC Med Ethics*, 4, *17*(1), 67.

Boddington, P., (2012). *Ethical challenges in genomics research*. Heidelberg: Springer.

Brakewood, B. I., & Poldrack, R. A. (2013). The ethics of secondary data analysis: Considering the application of Belmont principles to the sharing of neuroimaging data. *Neuroimage, 82*, 671–676.

Carter, P., Laurie, G. T., & Dixon-Woods, M. (2015). The social licence for research: Why care data ran into trouble. *Journal of Medical Ethics, 41*(5), 404–409.

Chadwick, R., & Berg, K. (2001). Solidarity and equity: New ethical frameworks for genetic databases. *Nature Reviews Genetics, 2*(4), 318–321.

Commission for the Study of Ethical Problems in Medicine and Biomedical and Behavioral Research. (1984). Ethics, politics, and access to health care: A critical analysis of the President's Commission for the Study of Ethical Problems in Medicine and Biomedical and Behavioral Research. *Cardozo Law Review, 6*(2), 303–320.

Contreras, J. L. (2011). Bermuda's legacy: Policy, patents, and the design of the Genome Commons. *Minnesota Journal of Law and Science & Technology, 12*, 61.

Currie, J. (2013). "Big Data" versus "Big Brother": On the appropriate use of large-scale data collections in pediatrics. *Pediatrics, 131*(Suppl 2), S127–S132.

Davenport, P., Cragg, W., Crago, M., Fanelli, D., Fleury, J.-M., Given, L. M., ... Stanton-Jean, M. (2010). *Honesty, accountability and trust: Fostering research integrity in Canada*. Ottawa: Council of Canadian Academies.

Dawson, A. (2010). The future of bioethics: three dogmas and a cup of hemlock. *Bioethics, 24*(5), 218–225.

Devinsky, O., Dilley, C., Ozery-Flato, M., Aharonov, R., Goldschmidt, Y., Rosen-Zvi, M., Clark, C., & Fritz, P. (2016a). Changing the approach to treatment choice in epilepsy using big data. *Epilepsy Behaviour, 56*, 32–37.

Devinsky, O., Hesdorffer, D. C., Thurman, D. J., Lhatoo, S., & Richerson, G. (2016b). Sudden unexpected death in epilepsy: Epidemiology, mechanisms, and prevention. *Lancet Neurology, 15*(10), 1075–1088.

Dove, E. S., Joly, Y., & Knoppers, B. M. (2012). Power to the people: A wiki-governance model for biobanks. *Genome Biology, 13*(5), 158.

Elliott, K. C., & Resnik, D. B. (2014). Science, policy, and the transparency of values. *Environmental Health Perspectives, 122*(7), 647–650.

Fischer, T., Brothers, K. B., Erdmann, P., & Langanke, M. (2016). Clinical decision-making and secondary findings in systems medicine. *BMC Medical Ethics, 17*(1), 32.

Fisher, L., Hessler, D. M., Polonsky, W. H., Masharani, U., Peters, A. L., Blumer, I., & Strycker, L. A. (2016). Prevalence of depression in Type 1 diabetes and the problem of over-diagnosis. *Diabetic Medicine, 33*(11), 1590–1597.

Gherardi, S., Østerlund, C. S., & Kensing, F. (2014). Editorial: Personal health records: Empowering patients through information systems. *Information Technology & People, 27*(4), 390–396.

Ginsburg, G. (2014). Medical genomics: Gather and use genetic data in health care. *Nature, 508*(7497), 451–453.

Global Network of WHO Collaborating Centres for Bioethics. (2015). *Global Health ethics: Key issues*. Luxembourg: WHO, Retrieved from http://www.who.int/ethics/publications/global-health-ethics/en/

Goodman, B. (2016). What's wrong with the right to genetic privacy: Beyond exceptionalism, parochialism and adventitious ethics. In B. D. Mittelstadt & L. Floridi (Eds.), *The Ethics of Biomedical Big Data* (pp. 139–167). Switzerland: Springer International Publishing.

Harris, J. R., Burton, P., Knoppers, B. M., Lindpaintner, K., Bledsoe, M., & Brookes, A. K. (2012). Towards a roadmap in global biobanking for health. *European Journal of Human Genetics, 20*, 1105–1111.

Hayden, E. C. (2012). *A broken contract*. London: Nature Publishing Group Macmillan Building. Retrieved from http://environmentportal.in/files/file/informed%20consent.pdf

Hays, R., & Daker-White, G. (2015). The care data consensus qualitative analysis of opinions expressed on Twitter. *BMC Public Health, 15*, 838.

Helgesson, G., (2012). In defense of broad consent. *Cambridge Quarterly of Health Ethics, 21*(1), 40–50.

Heubusch, K. (2006). Interoperability: What it means, why it matters. *Journal of AHIMA/ American Health Information Management Association, 77*(1), 26–30.

Hockings, E. (2016). A critical examination of policy – Developments in information governance and the biosciences. In B. D. Mittelstadt & L. Floridi (Eds.), *The Ethics of Biomedical Big Data* (pp. 95–118). Switzerland: Springer International Publishing.

Hoeyer, K. (2016). Denmark at a crossroad? Intensified data sourcing in a research radical country. In B. D. Mittelstadt & L. Floridi (Eds.), *The Ethics of Biomedical Big Data* (pp. 73–94). Switzerland: Springer International Publishing.

Hoffman, S. (2010). Electronic health records and research: Privacy versus scientific priorities. *American Journal of Bioethics, 10*(9), 19–20.

Hood, L., Lovejoy, J. C., & Price, N. D. (2015). Integrating big data and actionable health coaching to optimize wellness. *BMC Medicine, 13*(1), 4.

Hudson, K. L. (1995). Genetic discrimination and health insurance: An urgent need for reform. *Science, 270*(5235), 391–393.

Institute of Medicine. (2015). *Sharing clinical trial data: Maximizing benefits, minimizing risk.* Washington, DC: Institute of Medicine.

Karlsen, J. R., Solbakk, J. H., & Holm, S. (2011). Ethical endgames: Broad consent for narrow interests; open consent for closed minds. *Cambridge Quarterly of Healthc Ethics, 20*(4), 572–583.

Kaye, J. (2012). Embedding biobanks as tools for personalised medicine. *Norsk Epidemiology, 21*(2), 169–175.

Kaye, J., Whitley, E. A., Kanelloupoulou, N. K., Creese, S., Hughes, K. J., & Lund, D. (2011). Dynamic consent: A solution to a perennial, problem? *BMC, 343*, d6900–d6900.

Knoppers, B. M., & Chadwick, R. (2005). Human genetic research: Emerging trends in ethics. *Nature Review of Genetics, 6*(1), 75–79.

Kulynych, J., & Greely, H. T. (2017). Clinical genomics, big data, and electronic medical records: Reconciling patient rights with research when privacy and science collide. *Journal of Law and the Biosciences, 4*(1), 94–132.

Leonelli, S. (2014). What difference does quantity make? On the epistemology of big data in biology. *Big Data & Society, 1*(1). Doi: 10.1177/2053951714534395.

Lunshof, J. E., Chadwick, R., Vorhaus, D. B., & Church, G. M. (2008). From genetic privacy to open consent. *Nature Review of Genetics, 9*(5), 406–411.

Lupton, D. (2014). The commodification of patient opinion: The digital patient experience in age of big data. *Sociology of Health & Illness, 36*(6), 856–869.

Masic, I. (2012). Ethical aspects and dilemmas of preparing, writing and publishing of the scientific papers in the biomedical journals. *Acta Informatica Medica, 20*(3), 141–148.

Master, Z., Nelson, E., Murdoch, B., & Caulfield, T. (2012). Biobanks, consent and claims of consensus. *Nature Methods, 9*(9), 885–888.

Mayer-Schönberger, V., & Cukier, K. (2014). *Big data: A revolution that will transform how we live, work, and think.* Boston: Eamon Dolan/Mariner Books.

Morrison, M., Dickenson, D., & Lee, S. S. (2016). Introduction to the article collection "Translation in healthcare: ethical, legal, and social implications". *BMC Medical Ethics, 17*(1), 74.

OECD.(2013). The OECD privacy framework. Retrieved from http://www.oecd.org/internet/ieconomy/privacy-guidelines.htm

Prainsack, B., & Buyx, A. (2013). A solidarity-based approach to the governance of research biobanks. *Medical Law Review, 21*(1), 71–91.

Rabeharisoa, V., Callon, M., Filipe, A. M., Nunes, J. A., Paterson, F., & Vergnaud, F. (2014). From "politics of numbers" to "politics of singularisation": Patients' activism and engagement in research on rare diseases in France and Portugal. *BioSocieties, 9*(2), 194–217.

Rippen, H., & Risk, A. (2000). E-health code of ethics. *Journal of Medical Internet Research, 2*(2), e9.

Riso, B., Tupasela, A., Vears, D. F., Felzmann, H., Cockbain, J., Loi, M., ... Rakic, V. (2017). Ethical sharing of health data in online platforms – which values should be considered? *Life Science and Social Policy, 13*(1), 12.

Rodriguez, R. N. (2013). Building the big tent for statistics. *Journal American Statistical Association, 108*(501), 1–6.

Roski, J., Bo-Linn, G. W., & Andrews, T. A. (2014). Creating value in health care through big data: opportunities and policy implications. *Health Affairs, 33*(7), 1115–1122.

Rothstein, M. A., (2010). Is deidentification sufficient to protect health privacy in research? *American Journal of Bioethics, 10*(9), 3–11.

Salerno, J., Knoppers, B. M., Lee, L. M., Hlaing, W. M., & Goodman, K. W. (2017). Ethics, big data and computing in epidemiology and public health. *Annals of Epidemiology, 27*(5), 297–301.

Sane, J., & Edelsteinm, M. (2015). Centre on Global Health Security (Ed.), *Overcoming barriers to data sharing in public health. A global perspective*. London: Chatham House.

Schickhardt, C., Hosleym, N., & Winklerm, E. C. (2016). *Researchers' duty to share pre-publication data: From the prima facie duty to practice*. In B. D. Mittelstadt & L. Floridi (Eds.), *The Ethics of Biomedical Big Data* (pp. 309–338). Switzerland: Springer International Publishing.

Schuler, G. D., Boguski, M. S., Stewart, E. A., Stein, L. D., Gyapay, G., Rice, K. ... Hudson, T. J. (1996). A gene map of the human genome. *Science, 274*(5287), 540–546.

Shabani, M., & Borry, P. (2015). Challenges of web-based personal genomic data sharing. *Life Sciences, Society and Policy, 11*, 3.

Steinsbekk, K. S., Kåre Myskja, B., & Solberg, B. (2013). Broad consent versus dynamic consent in biobank research: Is passive participation an ethical problem? *European Journal of Human Genetics, 21*(9), 897–902.

Sterkckz, S. V., Rakic, J., & Cockbain, P. (2016). "You hoped we would sleep walk into accepting the collection of our data": Controversies surrounding the UK care. Data scheme and their wider relevance for biomedical research. *Medicine, Health Care and Philosophy, 19*(2), 177–190.

Suter, S. M. (2004). Disentangling privacy from property: Toward a deeper understanding of genetic privacy. *George Washington Law Review, 72*(4), 737–814.

ter Meulen, R. H. J., Newson, A. J., Kennedy, M. R., & Schofield, B. (2011). Background paper. *Genomics, health records, database linkage and privacy*. London: Nuffield Council on Bioethics.

United Nations Educational Social Cultural Organisation. (2003). *International declaration on human genetic data*. Retrieved from http://unesdoc.unesco.org/images/0013/001331/133171e.pdf#page=45

Vassilakopoulou, P., Skorve, E., & Aanestad, M. (2016). Premises for clinical genetics data governance: Grappling with diverse value logics. In B. D. Mittelstadt & L. Floridi (Eds.), *The Ethics of Biomedical Big Data* (pp. 239–256). Switzerland: Springer International Publishing.

Vayena, E., Salathe, M., Madoff, L. C., & Brownstein, J. S. (2015). Ethical challenges of big data in public health. *PLoS Computational Biology, 11*, e1003904.

Wellcome Trust. (2013). *Summary report of qualitative research into public attitudes to personal data and linking personal data*. Wellcome Trust. Retrieved from https://wellcomelibrary.org/item/b20997358#?c=0&m=0&s=0&cv=0

Whitley, E. A., Kanellopoulou, N., & Kaye, J. (2012). Consent and research governance in biobanks: evidence from focus groups with medical researchers. *Public Health Genome, 15*(5), 232–242.

SHAPING A CULTURE OF SAFETY AND SECURITY IN RESEARCH ON EMERGING TECHNOLOGIES: TIME TO MOVE BEYOND "SIMPLE COMPLIANCE" ETHICS

Monique Ischi and Johannes Rath

ABSTRACT

Most research ethic review procedures refer to the principles of safety and security only as sub-criteria of other ethical principles such as the protection of human subjects in research, thereby ignoring the public good aspect of safety and security. In addition, Research Ethics Review Committees (RECs) are usually dominated by philosophers, ethicists, medical doctors, and lawyers with limited practical backgrounds in safety and security risk management. This gap of knowledge restricts ethics reviews in carrying out project-specific safety and security risk management and defers this responsibility to lawmakers and national legal authorities. What might be sufficient in well-regulated and well-understood environments, such as the safety of individuals during clinical research, is insufficient in managing rapidly changing and emerging risks – such as with emerging biotechnologies – as well as addressing the public good dimension of safety and security.

This chapter considers governance approaches to safety and security in research. It concludes that legal mechanisms are insufficient to cope with the complexity of and the fast progress made in emerging technologies. The chapter also addresses the role and potential of research ethics as a safety and security governance approach. It concludes that research ethics can play an important role in the governance of such risks arising from emerging

Ethics and Integrity in Health and Life Sciences Research
Advances in Research Ethics and Integrity, Volume 4, 85–98

ISSN: 2398-6018/doi:10.1108/S2398-601820180000004006

technologies, for example through fundamental rights and public good considerations. However, in reality the current capacity of ethics in the safety and security governance of emerging technologies is limited. It is argued that in newly emerging areas of research currently applied legal compliance–based approaches are insufficient. Instead, inclusion of fundamental risk management knowledge and closer interactions between scientists, safety, and security experts are needed for effective risk management. Safety and Security Culture provide frameworks for such interactions and would well complement the current legal compliance–based governance approaches in research ethics.

Keywords: Virtue ethics; safety; security; dual use; ethics review; Horizon Europe; culture of ethics

INTRODUCTION

Current ethics governance approaches addressing safety and security tend to focus on a legal compliance approach. A practical example for such an approach, discussed later in this chapter, is the European Commission's (EC) Ethics Appraisal Procedure in Horizon 2020. The Ethics Appraisal Procedure uses a legal compliance–focused approach to address safety and security risks in research. We argue that this approach has resulted in limited success in addressing the complexity involved in governing safety and security issues related to emerging technologies. Whether future EC initiatives like Responsible Research and Innovation (RRI) will be more successful remains to be seen.

An alternative way forward in managing safety and security risks in emerging technologies is to move beyond legal compliance and REC-based governance approaches and focus on fostering a culture of safety and security, defined as the joint responsibility of all stakeholders in the research enterprise that is built on beliefs and attitudes and relies on well-informed educational programs. Individual initiatives in developing such a culture are briefly discussed in this chapter.

SAFETY AND ETHICS

Although research ethics rarely addresses safety as a stand-alone value or principle, the "condition of being safe from undergoing or causing hurt, injury or loss" (https://en.oxforddictionaries.com/definition/safety) lies at the heart of almost all ethical considerations in research. For example, individual human rights and human subject based safety concepts play a critical function in medical research ethics (Directive, 2001/20/EC). The step-wise design of clinical trials is first and foremost defined by individual safety considerations, ensuring that minimum safety risks apply to individual participants (Akbari et al., 2015; Champer, Buchman, & Akbari, 2016; Yao, Zhu, Jiang, & Xia, 2013).

Increasingly, safety is not only recognized as an individual right but also as a public good (Spiegel, 2003). Environmental ethics (http://www.iep.utm.edu/envi-eth/), for example, as an emerging ethics discipline, addresses safety as a public good (Harrison & Burgess, 2000), extending assessment scope and stakeholdership beyond any individuals directly affected. As such, restricting safety considerations to human subjects, as in medical research, is insufficient to address the full safety spectrum arising from emerging technologies.

SAFETY GOVERNANCE

Substantial global variation exists, when it comes to the governance of safety (Dabrowska, 2015), ranging from law-based and strictly enforced safety frameworks such as in the European Union (for example: https://osha.europa.eu/en/legislation/directives/exposure-to-chemical-agents-and-chemical-safety/) to, due to resources limitation, very loosely controlled safety provisions in some poorer countries (Mok, Gostin, Das Gupta, & Levin, 2010).

Even within the Western world, where a considerable part of research involving emerging technologies is conducted, substantial national differences exist. These differences are frequently caused by different legal traditions (Ashford, 2010) on the one hand and different risk management principles (Ashford, 2007) on the other. In the European Union and its civil law-based, positivist way of governance, a growing number of detailed laws have been passed over the last decades, addressing various kinds of safety risks (e.g., Occupational Health and Safety: https://osha.europa.eu/en/safety-and-health-legislation/european-directives, chemicals (http://eur-lex.europa.eu/legal-content/EN/TXT/?uri=CELEX:32008R1272, http://eur-lex.europa.eu/legal-content/EN/TXT/?uri=CELEX:32008R1272), radioactivity (https://ec.europa.eu/energy/en/overview-eu-radiation-protection-legislation), and genetically modified organisms (GMOs) (https://ec.europa.eu/food/plant/gmo/legislation_en). Common to all these regulations is the focus on structured, *ex-ante* risk assessments in defining correct levels of risk mitigation. Even, if parts of the risk assessment are done in a way of self-governance (e.g., biosafety risk level definition in GMOs), high-risk experiments usually require prior authorization by public institutions. In such a framework, safety is quickly perceived as a purely legal/regulatory compliance issue, where the key responsibility in risk management rests on lawmakers or administrative regulators. The main responsibility left to the researchers is to show how they intend to comply with this framework and to obtain the relevant authorizations as well as implementing the requirements outlined in these authorizations when actually conducting the research.

Although such legal/regulatory compliance approach may be suitable for solving standard methodological safety questions in well-defined risk environments of research (e.g., working with radioactive materials or well-defined chemicals), it is incapable of mitigating against emerging risks that are not already regulated at the legal level. Since the devising, adoption, and application of laws often takes years, if not decades, many of the risks currently discussed in the context of emerging technologies are not at all, inadequately or only partially, regulated by law (Sandler, 2014). This delay in the implementation of legislation constitutes a serious challenge to legal/regulatory compliance–based governance frameworks in safety and security.

In contrast, safety governance in the United States (US) is more complex and involves legislative and non-legislative approaches (http://www.triumvirate.com/state-by-state-biosafety-regulations). A distinguishing feature of the US, differing from the European research situation, is the stronger reliance on self-governance with regard to safety that often builds on non-legally binding safety guidelines for research (https://www.cdc.gov/biosafety/publications/bmbl5/BMBL.pdf). Another difference is that the motivation for adhering to relevant guidelines may arise from stronger liability provisions in case of safety breaches (Attaran & Wilson, 2015; Avilds, 2014; Glass & Freedman, 1991). Recent biosafety incidents in the US seem to indicate, however, that there are substantial weaknesses in self-governance and compliance (Baskin, Gatter, Campbell, Dubois, & Waits, 2016; Weiss, Yitzhaki, & Shapira., 2015) which have not only resulted in dangerous incidents but also triggered a government review of current practices (Kaiser, 2014).

The effectiveness of soft-law approaches based on self-governance will highly depend on behavioral elements like the willingness of individuals and institutions to adhere to non-binding guidelines. Consequently, US efforts in improving safety in research have for a long time focused on concepts of individual accountability within the framework of responsible science (Federman, Hanna, & Rodriguez, 2002), rather than on legal compliance. Initiatives and new concepts to foster a culture of responsibility regarding safety are needed (Berkelman & Le Duc, 2014) to overcome challenges to such self-governance-based approaches (e.g., conflicts of interest both on an individual but also on an organizational level, standardization).

A different safety governance problem is related to uncertainty (Savage, Gorman, & Street, 2013). Because of the novelty of the technologies and methodologies, knowledge about the nature or extent of risks does not exist, making uncertainty a key risk management challenge in emerging technologies. To account for uncertainty in safety risk management, EU law usually refers to the Precautionary Principle (http://eur-lex.europa.eu/legal-content/EN/TXT/?uri=URISERV%3Al32042). A consequent interpretation of the Precautionary Principle results in a reversal of the burden of proof (Goldstein & Carruth, 2004), whereby the potential creator of a risk has to prove that the technology is safe and no unacceptable risks are introduced. Proving the non-existence of anything is inherently difficult.

Furthermore, different interpretations of the principle exist (Conko, 2003), resulting in inconsistencies not only at the global level (Johnson & Casagrande, 2016) but also among EU countries in its application (Garnett & Parsons, 2017; Rath, Ischi, & Perkins, 2014). The area of green biotechnology (https://en.wikipedia.org/wiki/Biotechnology) has been the key showcase for the (at times) irrational discussions that emerge because of applying the same Precautionary Principle concept in different safety and national contexts (Rogers, 2017; Tagliabue, 2016).

Thus, relying on a legal compliance–based framework centered on the Precautionary Principle, to address safety questions in emerging areas of research faces serious challenges of consistency and introduces substantial risks to the research enterprise as a whole by being overly prohibitive. Globally accepted versions of the Precautionary Principle are urgently needed that allow the benefits of research to fully materialize.

Research ethics through a REC engages in risk management of emerging technologies in the initial stages of the research cycle. At these stages, the existences as well as the magnitudes of risks arising from emerging technologies are often uncertain. "Horizon scanning" (https://ec.europa.eu/research/foresight/index.cfm?pg=horizon) is frequently applied to identify risks and benefits (O'Malley & Jordan, 2009) but hardly ever provides more than scenarios and often disproportionately focuses on unlikely worst- and best-case scenarios (Gordijn, 2005). Referring to the Precautionary Principle to mitigate such highly unlikely extreme worst-case scenarios can result in highly prohibitive outcomes and negative implications for the advancement of science and the freedom of research (Sunstein, 2002). Applying the Precautionary Principle at this early stage of research might even impact negatively on safety itself, by restricting the use of such novel technologies to overcome existing safety concerns and public health risks (Duprex, Fouchier, Imperiale, Lipsitch, & Relman, 2015). As such, transferring the Precautionary Principle as a concept for risk management outlined in OSHA, consumer protection, and environmental safety legislation to risk assessments in emerging technology research needs modifications. For example, the "proportionality" sub-test of the Precautionary Principle (http://eur-lex.europa.eu/legal-content/EN/TXT/?uri=URISERV%3Al32042) needs to address also the negative implications on the freedom of and progress in science and long-term negative safety implications by not developing new potentially safer technologies and allowing research progress within controlled environments to gain knowledge and reduce uncertainty over risks.

USING ETHICS REVIEW TO GOVERN SAFETY IN EMERGING TECHNOLOGIES

A concrete example, where ethics governance is used to govern the safety of emerging technologies, is the Ethics Appraisal Procedure (http://ec.europa.eu/research/participants/docs/h2020-funding-guide/cross-cutting-issues/ethics_en.htm) carried out by the European Commission (EC) in its research Framework Project Horizon 2020. Although the Ethics Appraisal Procedure includes a self-assessment step, it addresses safety primarily as legal compliance concern, which in practice means the use of a checklist for ensuring compliance of European and/or national legislation by requesting copies of relevant approvals (http://ec.europa.eu/research/participants/data/ref/h2020/grants_manual/hi/ethics/h2020_hi_ethics-self-assess_en.pdf). The reasons for restricting risk management to this basic and, as argued earlier, often inadequate level are manifold. One reason seems to be the varying safety expertise of ethics panel members which limits the capacity of such panels to carry out independent expert safety assessments. However, restricting the Ethics Appraisal Procedure to a legal compliance check in safety is problematic for four reasons:

First, due to the novelty of the technologies, uncertainty may exist over whether existing laws are applicable. As an example for this kind of uncertainty, the European Commission (EC) at the time of writing this chapter has not defined whether organisms manipulated using genome editing technologies (e.g., CRISPR/Cas9) qualify as genetically modified organisms. This creates uncertainty over the applicability of relevant legislation regarding genetically modified organisms. Since the Ethics Appraisal Procedure addresses safety through a legal compliance concept, it creates substantial uncertainty over what to do.

Second, applying a Precautionary Principle interpretation to research that has been developed in EU law for applications outside research limits the chances for potentially revolutionizing technologies to be funded. Research involving gain of function experiments in pathogens is such an example for this situation, where uncertainty over the outcome of the mutation-selection experiment has led to a moratorium (Malakoff, 2013; https://www.phe.gov/s3/dualuse/Documents/gain-of-function.pdf) and/or the upgrade of biosafety levels. Following the logic of the Precautionary Principle and limiting such research to Biosafety Level 4 (http://dip21.bundestag.de/dip21/btd/18/013/1801380.pdf) or completely banning such experiments (Reardon, 2014) is highly prohibitive as only a limited number of laboratories support BSL 4 (https://en.wikipedia.org/wiki/Biosafety_level). Alternative ways like the focus on technical safeguards (e.g., molecular containment, Baas, 2013) are not within the remit of a solely legal/regulatory compliance-based Ethics Appraisal Procedure, but would mobilize the innovative potential of the scientific community to foster safety as well as reduce uncertainty and risks.

Third, due to various legal interpretations of the Precautionary Principle, inconsistencies in biosafety requirements exist. For example, the mentioned H5N1 gain of function experiments can only be carried out at BSL 4 in Germany, whereas in the Netherlands such experiments can also be done at BSL 3 (Herfst et al., 2012). This may result in "approval shopping" and ethics dumping at the costs of safety as well as discrimination of researchers within the same funding scheme.

Fourth, if appraisals are strictly limited to the question of whether and what national permits are needed; there is very limited added value for such an appraisal. Compliance with national legislation is best assessed and enforced through national regulatory authorities and not international RECs.

None of these outcomes is desirable and the current approach taken by the EC in paralleling ethics with legal compliance when it comes to the safety and security governance of emerging technologies should be reassessed.

SECURITY AND ETHICS

Security is in the eye of the beholder and "the state of being free from danger or threat" (https://en.oxforddictionaries.com/definition/security) can apply to different entities (e.g., individuals or collectives) as well as to different interests (e.g. national security, civil security, human rights; https://en.wikipedia.org/wiki/Security). As a consequence, several security concepts exist, which may overlap at times (e.g., civil security and human security) but sometimes even compete (such as in the case of national security and human security). Therefore, a clear distinction between these different framings of security is warranted before engaging into any ethics or morality discussion (Rath et al., 2014).

SECURITY GOVERNANCE

As with the concept of safety, substantial national differences exist when it comes to the governance of security risks in research. As an example, emerging biotechnologies in the US may be directly affected by the Select Agent Regulation (https://www.selectagents.gov/regulations.html), if such experiments involve pathogens listed in the Regulation. In contrast, no such legislation exists in the European Union (EU), although similar legislation may exist in individual EU member states (http://www.legislation.gov.uk/ukpga/2001/24/schedule/5). The implementation of the Select Agent Regulation faces challenges related to sometimes arbitrary inclusion and exclusion criteria for pathogens (Enserink & Malakoff, 2001) but also in being overly restrictive and destructive on much-needed research (Casadevall & Imperiale, 2010; Matthews, 2012). Missing global harmonization on such legislation also limits its impact on security at the national and international scale.

International export control regimes (http://www.australiagroup.net/en/, http://www.wassenaar.org/, http://mtcr.info/, http://www.un.org/en/sc/1540/) have created a more homogenous environment in security governance between the US and Europe. Export control legislation governs the transboundary movement of certain materials, agents, technologies, and knowledge that could be used, for example, in bioweapons development or production (Shaw, 2016, http://ec.europa.eu/trade/import-and-export-rules/export-from-eu/dual-use-controls/, https://www.state.gov/strategictrade/overview/). Despite this international harmonization, there is often uncertainty over whether such export control legislation is actually applicable to emerging technologies and research.

First, export control legislation frequently contains a "catch-all-clause" to ensure coverage of all potentially relevant agents, technologies, and information (https://www.state.gov/strategictrade/practices/c43179.htm). This creates uncertainties when defining the exact inclusion criteria of the legislation. Furthermore, varying interpretations of this clause by countries create additional challenges (Stewart & Bauer, 2015). Second, export control legislations usually contain exemptions for basic or fundamental research. The remits of these exemptions are not clear and recent rulings by national courts have added more to the confusion as to what extent such exemptions exist (Enserink, 2013; Enserink, 2015).

Thus, security governance of emerging technologies outside export control legislation is often nonexistent, and even within the limited protection provided by export control legislation, exemptions may lead to the nonapplicability of such legislation in governing security risks arising from emerging technologies.

USING ETHICS REVIEW TO GOVERN SECURITY IN EMERGING TECHNOLOGIES

When it comes to ethics as a governance tool for security only a few practical examples exist with the Horizon 2020 Ethics Appraisal Procedure implemented by the EC being one of the more advanced frameworks. The Ethics Appraisal Procedure builds on expert expertise recruited in the review panels to provide advice to the EC and define requirements for the individual research projects that then become contractual obligations. There are several lessons to be learnt from the participation in these panels in the context of security.

First, security is not a coherent concept. Today, the Ethics Appraisal Procedure, for example, addresses security through three different framings (dual use, misuse, and exclusive civilian application focus). Dual use is addressed in the context of export control legislation (Council Regulation (EC) No 428/2009 of 5 May 2009 setting up a Community regime for the control of exports, transfer, brokering and transit of dual-use items (Recast) http://eur-lex.europa.eu/legal-content/EN/TXT/?qid=1489490204085&uri=CELEX:02009R0428-20161116), which combines military, national, civil, and human security interests. Misuse addresses specific civil security issues related to terrorism (http://ec.europa.eu/research/participants/data/ref/h2020/other/hi/guide_research-misuse_en.pdf). Finally, the criterion of "civilian application focus" (http://ec.europa.eu/research/participants/data/ref/h2020/other/hi/guide_research-civil-apps_en.pdf) takes a very strong human security position by restricting funding for the military and the development of weapons.

Second, when it comes to security and ethics, human security-based concepts (https://en.wikipedia.org/wiki/Human_security) with their many overlaps with human rights are easier to integrate in ethics deliberations. Practical examples of integrating human security-based approaches into ethics have become evident in public discussions on human rights-sensitive security research projects (Rath et al., 2014). Within the area of emerging technologies, the H5N1 controversy provides an example where health security as one element of human security and fundamental rights (e.g. freedom of expression, information and science) on one side conflict with national security interests (Potential Risks and Benefits of Gain-of-Function Research: Summary of a Workshop Board on Life Sciences; Division on Earth and Life Studies; Committee on Science, Technology, and Law; Policy and Global Affairs; Board on Health Sciences Policy; National Research Council; Institute of Medicine. Washington (DC): National Academies Press (US); 2015 Apr 13. ISBN-13: 978-0-309-36783-7ISBN-10: 0-309-36783-2) on the other. Attempts to limit the distribution of the research results through various means, including export control measures, conflicted with human security and rights interests, because the dissemination of knowledge on what defines pathogenicity and transmissibility of a virus supports the understanding of disease mechanisms and in the long term can contribute to the development of adequate counter measures

Third, the Horizon 2020 Ethics Appraisal Procedure also provides for interesting lessons regarding the interplay between law and other governance approaches, including ethics. While the US approach focused on a nonlegal approach through scientific advisory boards and funding restrictions, the Netherlands and the EC applied a two-tier approach in the H5N1 discussion. On the one hand, the Netherlands followed a legal approach and invoked export control license requirements for the "export" of information. On the other hand, an Ethics Audit took place at the EC of EC-funded gain of function research which was used to introduce additional safeguards through contractual obligations in order to ensure safety and security.

Fourth, only a very limited number of people with practical expertise in research, security, law, and ethics are available to carry out such appraisals. The quality of the assessments will depend on the recruitment of this expertise into the panels. Access and inclusion of high-level expertise in security issues related to emerging technologies have proven to be a severely limiting factor in ensuring high-quality security risk management.

WHERE DO WE STAND – WHERE SHOULD WE GO

As outlined before, the role that ethics and law play in the governance of safety and security of emerging technologies varies substantially between countries. For example, safety issues in the European Union are frequently addressed at the legal level and the role of ethics is seen as ensuring compliance to legal requirements. This creates problems when it comes to unregulated areas (e.g., genome editing) where a compliance approach is insufficient since no legislation exists that would address the specific safety challenges of the new technology. Without high-level safety risk management expertise in RECs, addressing legally unregulated complex safety issues becomes technically impossible. The same argument is valid for security as only a very limited and inconsistent number of legal standards exists outside the export control area that would be applicable to research in emerging technologies. Adjustments in the composition of RECs to account for such expertise in safety and security, however, are rarely done and therefore current ethics review concepts are inadequate risk management frameworks for the safety and security risks arising from emerging technologies.

In contrast, in the United States safety governance is focused on guidance documents and compliance usually not legally enforced outside the public funding, public laboratory, and Select Agent Regulation area. Indirect enforcement of safety guidelines might also be driven by liability concerns.

Guidance documents, in contrast to laws and decrees, are usually more generic and risk management becomes less standardized, more flexible, and better suitable to adjust to unforeseen challenges. The efficiency of this soft-law approach, however, will depend on the willingness of the individual researcher and organization to engage, adhere, and devote resources to this, and current biosafety incidents put in question whether this approach is sufficiently robust, or will need re-adjustments.

Given this situation, it is clear that at the current time there does not exist a functional risk management approach to safety and security in research in general. New and unprecedented safety and security risks arising from the development of emerging technologies are further sharpening an already unsatisfactory situation. What is needed are organizational frameworks that foster the acquisition, development, tailoring, and integration of safety and security risk management knowledge into the technology development process with clear responsibilities and accountabilities for all stakeholders involved. Organizational frameworks that foster all of these elements should be adopted by research institutions. An established framework in this context is *Safety and Security Culture* (https://www.iaea.org/topics/safety-and-security-culture).

THE ADDED VALUE OF *SAFETY AND SECURITY CULTURE*

Safety and Security Culture is built on an organizational culture concept (Schein & Schein, 2017). The International Atomic Energy Agency (IAEA), for example, defines safety culture as "the assembly of characteristics and attitudes in organizations and individuals which establishes that, as an overriding priority, protection and safety issues receive the attention warranted by their significance" (http://www-ns.iaea.org/tech-areas/operational-safety/safety-culture-home.asp).

Creating a culture of safety and security that goes beyond legal compliance and includes the various stakeholders of the scientific enterprise will be critical to manage currently unregulated research activities in the area of emerging technologies. Although top-down approaches (e.g., legislative acts) might have an impact in shaping cultures (e.g., in assigning accountabilities and liabilities), most of the risk management will require an inclusive bottom-up approach involving researchers, educators, academic and research institutions, funders, as well as safety and security experts.

THE POTENTIAL FOR *SAFETY AND SECURITY CULTURE* IN THE CONTEXT OF FRAMEWORK PROGRAM 9 (FP9)

At the time of writing, the European Commission has started discussing the next Framework Program called Framework Program 9 (FP9). These large FPs offer unique opportunities to develop and implement research standards across and beyond the European Union. FP9 focuses on the concept of Responsible Research and Innovation (RRI) to manage the societal implications of research (http://ec.europa.eu/research/evaluations/pdf/hlg_issue_papers.pdf#view=fit&pagemode=none). In contrast to existing US responsibility frameworks (Responsible Research with Biological Select Agents and Toxins (2009) National Research Council (US) Committee on Laboratory Security and Personnel Reliability Assurance Systems for Laboratories Conducting Research on Biological Select Agents and Toxins. Washington (DC): National Academies Press (US);ISBN-13: 978-0-309-14535-0ISBN-10: 0-309-14535-X) and international (WHO: Responsible life sciences research for global health security A guidance

Document (2010) http://apps.who.int/iris/bitstream/10665/70507/1/WHO_HSE_GAR_BDP_2010.2_eng.pdf), RRI does not address safety and security explicitly (https://ec.europa.eu/programmes/horizon2020/en/h2020-section/responsible-research-innovation), and the risks are high that FP9 will have no effective safety and security risk management framework in place.

Safety and Security Culture can provide an alternative risk management framework by focusing on the human factor within a comprehensive framework of stakeholders. Attitudes, behaviors, and knowledge of individuals become key assets in increasing safety and security. Shaping them through education must be a key priority. Building up the expertise among young scientists in managing the risks to safety and security should become a critical long-term strategy. For the time being there is a critical shortage in education and university courses on safety and security risk management in emerging technologies.

Stakeholder engagement (e.g., research institutions, funders, lawmakers) in safety and security will most likely continue to be event driven, meaning that safety and security incidents or new legal requirements will shape institutional policies. The willingness to contribute and improve *Safety and Security Culture* within the research enterprise will therefore primarily be ad hoc and reactive.

Successful initiatives of *Safety and Security Culture* exist. The IAEA framework for safety and security culture in the nuclear area is maybe the most advanced and institutionalized version and is part of international treaties. Safety culture in the context of chemistry has been promoted by the American Chemical Society (Bertozzi, 2016, https://www.acs.org/content/acs/en/education/students/graduate/creating-safety-cultures-in-academic-institutions.html, https://www.acs.org/content/acs/en/chemical-safety/guidelines-for-chemical-laboratory-safety/resources-supporting-guidelines-for-chemical-laboratory-safety/building-strong-safety-cultures.html). A more comprehensive approach to CBRN Security Culture was discussed in a series of workshops under the United Nations Security Council Resolution 1540 (Thompson & Gahlaut, 2015) and European policymakers (https://ec.europa.eu/home-affairs/sites/homeaffairs/files/what-we-do/policies/pdf/com_2009_0273_en.pdf). Murphy (2012) provides an overview on safety culture initiatives in the biotechnology area.

CONCLUSIONS

Current risk management approaches in addressing safety and security risks from emerging technologies are inadequate. The complexity and uncertainty of safety and security issues in emerging technologies require new comprehensive risk management frameworks that focus on the human factor. The human subject thereby plays a critical role not only in vigilance and stewardship but also in knowledge generation and distribution. The concept of a *Safety and Security Culture* provides a promising novel approach to the safety and security challenges posed by emerging technologies. Some of the potential outcomes of emerging technologies have been identified by scholars as potentially existential threats to the human race (http://www.wired.co.uk/article/10-threats-civilisation-ai-asteroid-tyrannical-leader). Therefore, investments in the safety and security of these technologies are highly warranted.

REFERENCES

Akbari, O. S., Bellen, H. J., Bier, E., Bullock, S. L., Burt, A., Church, G. M., … Wildonger, J. (2015, Aug 28). Safeguarding gene drive experiments in the laboratory. *Science*, *349*(6251), 927–929. doi:10.1126/science.aac7932.

Ashford, N. (2007). The legacy of the precautionary principle in US law: The rise of cost-benefit analysis and risk assessment as undermining factors in health, safety and environmental protection. Retrieved from https://ashford.mit.edu/sites/default/files/documents/C28.%20LegacyOf Precaution_19.pdf

Ashford, N. (2010). Reflections on environmental liability schemes in the United States and European Union: Limitations and prospects for improvement. Retrieved from https://dspace.mit.edu/bitstream/handle/1721.1/55293/reflections_environmental-liability-schemes_us-eu.pdf?sequence=1

Attaran, A., & Wilson, K. (2015, Dec 1). The Ebola vaccine, iatrogenic injuries, and legal liability. *PLoS medicine*, *12*(12), e1001911. doi:10.1371/journal.pmed.1001911.

Avilds, M. R.. (2014, Dec). Compensation of research-related injuries in the European Union. *European Journal of Health Law*, *21*(5), 473–487.

Baas, T.. (2013). Molecular biocontainment for the flu. *Science-Business eXchange*, *6*(35), 949. doi:10.1038/scibx.2013.949.

Baskin, C. R., Gatter, R. A., Campbell, M. J., Dubois, J. M., & Waits, A. C. (2016). Self-regulation of Science: What can we still learn from Asilomar? *Perspectives in Biology and Medicine*, *59*(3), 364–381. doi:10.1353/pbm.2016.0031.

Berkelman, R. L., & Le Duc, J. W.. (2014). Culture of responsibility. *Science*, *345*(6201), 1101.

Bertozzi, C. R.. (2016, Nov 23). Ingredients for a positive safety culture. *ACS Central Science*, *2*(11), 764–766. doi:10.1021/acscentsci.6b00341.

Casadevall, A., & Imperiale, M. J.. (2010, Jun). Destruction of microbial collections in response to select agent and toxin list regulations. *Biosecurity and bioterrorism: biodefense strategy, practice, and science*, *8*(2), 151–154. doi:10.1089/bsp.2010.0012.

Champer, J., Buchman, A., & Akbari, O. S. (2016). Cheating evolution: Engineering gene drives to manipulate the fate of wild populations. *Nature Reviews Genetics*, *17*(3), 146–159. doi:10.1038/nrg.2015.34.

Conko, G. (2003). Safety, risk and the precautionary principle: rethinking precautionary approaches to the regulation of transgenic plants. *Transgenic Research*, *12*(6), 639–647.

Dabrowska, P. (2015). *Essays on global safety governance: challenges and solutions*. Warsaw: Centre for Europe, University of Warsaw. Retrieved from https://osha.europa.eu/en/legislation/directives/exposure-to-chemical-agents-and-chemical-safety/

Directive 2001/20/EC. OF THE EUROPEAN PARLIAMENT AND OF THE COUNCIL of 4 April 2001 on the approximation of the laws, regulations and administrative provisions of the Member States relating to the implementation of good clinical practice in the conduct of clinical trials on medicinal products for human use. OJ L 121, 1.5.2001, p. 34–44.

Duprex, W. P., Fouchier, R. A., Imperiale, M. J., Lipsitch, M., & Relman, D. A.. (2015, Jan). Gain-of-function experiments: time for a real debate. *Nature reviews. Microbiology*, *13*(1), 58–64. doi:10.1038/nrmicro3405.

Enserink, M. (2013). Flu researcher Ron Fouchier loses legal fight over H5N1 studies. Retrieved from http://www.sciencemag.org/news/2013/09/flu-researcher-ron-fouchier-loses-legal-fight-over-h5n1-studies

Enserink, M. (2015). Dutch appeals court dodges decision on hotly debated H5N1 papers. Retrieved from http://www.sciencemag.org/news/2015/07/dutch-appeals-court-dodges-decision-hotly-debated-h5n1-papers

Enserink, M., & Malakoff, D.. (2001). Congress weighs select agent update. *Science*, *294*(5546), 1438. doi:10.1126/science.294.5546.1438a.

Federman, D. D., Hanna K. E., & Rodriguez, L. L. (2002). *Responsible research: A systems approach to protecting research participants*. The National Academies Collection. Retrieved from https://www.ncbi.nlm.nih.gov/books/NBK43563/pdf/Bookshelf_NBK43563.pdf

Garnett, K., & Parsons, D. J.. (2017, Mar). Multi-case review of the application of the precautionary principle in European Union law and case law. *Risk Analysis*, *37*(3), 502–516. doi:10.1111/risa.12633.

Glass, K. C., & Freedman, B. (1991, Apr). Legal liability for injury to research subjects. *Clinical and Investigative Medicine*, *14*(2), 176–180.

Goldstein, B. D., & Carruth, R. S.. (2004). Implications of the precautionary principle: Is it a threat to science? *International Journal of Occupational Medicine and Environmental Health*, *17*(1), 153–161.

Gordijn, B. (2005, Oct). Nanoethics: from utopian dreams and apocalyptic nightmares towards a more balanced view. *Science and engineering ethics*, *11*(4), 521–533.

Harrison, C., & Burgess. (2000). Valuing nature in context: the contribution of common-good approaches. *Journal of Biodiversity & Conservation*, *9*(8), 1115–1130 doi:10.1023/A:1008930922198.

Herfst, S., Schrauwen, E. J., Linster, M., Chutinimitkul, S., de Wit, E., Munster, V. J., ... Fouchier, R. A. (2012, Jun 22). Airborne transmission of influenza A/H5N1 virus between ferrets. *Science*, *336*(6088), 1534–1541. doi:10.1126/science.1213362.

Johnson, B., & Casagrande, R.. (2016). Comparison of international guidance for biosafety regarding work conducted at biosafety level 3 (BSL-3) and gain-of-function (GOF) experiments. *Applied Biosafety*, *21*(3), 128–141, doi:10.1177/1535676016661772.

Kaiser, J. (2014). Lab incidents lead to safety crackdown at CDC. Retrieved from http://www.sciencemag.org/news/2014/07/lab-incidents-lead-safety-crackdown-cdc

Malakoff, D. (2013). H5N1 researchers announce end of research moratorium. Retrieved from http://www.sciencemag.org/news/2013/01/h5n1-researchers-announce-end-research-moratorium.

Matthews, S. (2012, Dec). Select-agent status could slow development of anti-SARS therapies. *Nature Medicine*, *18*(12), 1722. doi:10.1038/nm1212-1722.

Mok, E. A., Gostin, L. O., Das Gupta, M., & Levin, M.. (2010). Implementing public health regulations in developing countries: Lessons from the OECD countries. *The Journal of Law, Medicine & Ethics*, *38*(3), 508–519.

Murphy, J. F. (2012). Safety considerations in the chemical process industries. In J. A. Kent (Ed.), *Handbook of industrial chemistry and biotechnology* (pp. 47–105). New York: Springer.

O'Malley, S. P., & Jordan, E.. (2009, Jul). Horizon scanning of new and emerging medical technology in Australia: Its relevance to Medical Services Advisory Committee health technology assessments and public funding. *International Journal of Technology Assessment in Health Care*, *25*(3), 374–382. doi:10.1017/S0266462309990031.

Rath, J., Ischi, M., & Perkins, D. (2014). Evolutions of different dual-use concepts in international and national law and its implications on research ethics and governance. *Science and Engineering Ethics*, *20*(3), 769–790.

Reardon, S. (2014). US suspends risky disease research: Government to cease funding gain-of-function studies that make viruses more dangerous, pending a safety assessment. *Nature*, *514*, 411–412. Retrieved from http://www.nature.com/news/us-suspends-risky-disease-research-1.1619

Rogers, A. (2017). Who wants disease-resistant gm tomatoes? Probably not Europe. Retrieved from https://www.wired.com/2017/05/wants-disease-resistant-gm-tomatoes-probably-not-europe/

Sandler, R. (2014). *Ethics and emerging technologies*. New York: Palgrave Macmillan.

Savage, N., Gorman, M. E., & Street, A. (2013). *Emerging technologies: Socio-behavioral life cycle approaches*. Boca Raton, FL: CRC Press, ISBN 9814411000, 9789814411004.

Schein, E. H., & Schein, P. (2017). *Organizational culture and leadership* (5th ed.). Hoboken, NJ: John Wiley & Sons.

Shaw, R.. (2016, Apr). Export controls and the life sciences: controversy or opportunity? Innovations in the life sciences' approach to export control suggest there are ways to disrupt biological weapons development by rogue states and terrorist groups without impeding research. *EMBO Reports*, *17*(4), 474–480.

Spiegel, M. (2003). Public safety as a public good. In markets, pricing, and deregulation of utilities. *Topics in Regulatory Economics and Policy Series Book Series*, *40*, 183–200.

Stewart, I. J., & Bauer, S. (2015). Dual use export controls. Workshop Report, European Parliament, Directorate for External Policies. Retrieved from http://www.europarl.europa.eu/RegData/etudes/STUD/2015/535000/EXPO_STU%282015%29535000_EN.pdf

Sunstein, C. R. (2002). Beyond the precautionary principle. In J. M. Olin (Ed.), *Program in law and economics*. Working Paper No. 149. Retrieved from https://chicagounbound.uchicago.edu/cgi/viewcontent.cgi?article=1086&context=law_and_economics

Tagliabue, G.. (2016, Jun 25). The Precautionary principle: Its misunderstandings and misuses in relation to "GMOs". *New biotechnology*, *33*(4), 437–439. doi:10.1016/j.nbt.2016.02.007.

Thompson, J., & Gahlaut, S. (Eds.). (2015). *CBRN security culture in practice Editors* (Vol. 121). Amsterdam: IOS Press. ISBN: 978-1-61499-533-3 (print); 978-1-61499-534-0 (online).

Weiss, S., Yitzhaki, S., & Shapira. (2015). Lessons to be learned from recent biosafety incidents in the United States. *The Israel Medical Association Journal: IMAJ*, *17*(5), 269–273.

Yao, B., Zhu, L., Jiang, Q., & Xia, H. A. (2013). Safety monitoring in clinical trials. *Pharmaceutics*, *5*(1), 94–106. doi:10.3390/pharmaceutics5010094.

GOVERNING GENE EDITING IN THE EUROPEAN UNION: LEGAL AND ETHICAL CONSIDERATIONS

Mihalis Kritikos

ABSTRACT

The chapter analyses the re-emergence of gene editing as an object of policy attention at the European Union (EU) level. Editing the genome of plants and/or animals has been a rather controversial component of all EU policies on agricultural biotechnology since the late 1980s. The chapter examines in detail the various initiatives that have been assumed for the regulation of gene editing at the EU level. Since the first political and legislative attempts, the field has been revolutionized with the development of the CRISPR-Cas9 system, which is comparatively much easier to design, produce, and use. Beyond the pure, safety-driven scientific questions, gene editing, in its contemporary form, raises a series of ethical and regulatory questions that are discussed in the context of the legal options and competences of the EU legislators. Special attention is paid to questions about the legal status of gene editing in Europe and the adequacy of the current GMO framework to deal with all the challenges associated with the latest scientific developments in the field of gene editing with a special focus on gene drive. Given the ongoing discussions regarding the ethical tenets of gene editing, the chapter investigates the question on whether there is a need to shape an EU-wide "intervention" that will address the complex and dynamic socio-ethical challenges of gene editing and puts forward a series of proposals for the framing of an inclusive framework that will be based on the need to re-enforce public trust in the EU governance of emerging technologies.

Keywords: CRISPR-Cas9; gene editing; gene drive; EU law; ethics; risk assessment; proceduralism

Ethics and Integrity in Health and Life Sciences Research
Advances in Research Ethics and Integrity, Volume 4, 99–114

ISSN: 2398-6018/doi:10.1108/S2398-601820180000004007

INTRODUCTION

This chapter analyses the re-emergence of gene editing as an object of policy attention at the European Union (EU) level. Editing the genome of plants and/or animals has been a rather controversial component of all EU policies on agricultural biotechnology since the late 1980s. The first generation of genome editing tools appeared some 20 years ago in the regulatory agendas of EU law-makers. Since then, the field has been revolutionized with the development of the CRISPR-Cas9 system, which is comparatively much easier to design, produce and use. The CRISPR-Cas9 system currently stands out as the fastest, cheapest and most reliable system for "editing" genes. It is seen as the most significant game changer in the field of gene editing, due to its high degree of reliability and effectiveness, as well as its low cost.

Assessing and authorizing the commercial application of this technology has been a contentious political exercise that has, in effect, questioned the legitimacy of the EU institutional machinery to regulate such value-laden, risk-related technologies. As a result, the current EU legislative framework that focuses on the licensing of agricultural biotechnology has been applied unevenly across Europe, leading to major political conflicts and, eventually, to the first-ever re-nationalization of EU competences. The latter was introduced through the adoption of new EU rules allowing Member States to prohibit or restrict the cultivation of crops containing genetically modified organisms (GMOs) on their territory: even if they are permitted at the EU level.[1]

In other words, recent developments in genome editing, including the 2015 report of the US National Academy of Sciences (NAS), US National Academy of Sciences (2015), Kaiser (2017) and Reardon (2016), the 2015 Statement on genome editing technologies of the Council of Europe (Council of Europe, 2015), the 2015 statement of the Hinxton Group (The Hinxton Group, 2015), in particular through the use of CRISPR-Cas9 and the continuing emergence of novel applications and innovative methods, seem to bring back onto the EU discussion table a series of important questions. These questions refer to whether gene editing in its latest versions should be regulated or controlled at a transnational level; whether a process-based or product-based approach should be followed in that respect; and what is the role of the precautionary principle, including the plausibility and legitimacy of a moratorium (Bosley et al., 2015), and the terms of interface between scientific and ethical considerations when designing a governing framework for these rapidly evolving technical developments (Baltimore, Berg, & Botchan, 2015). Compared to the challenges brought forward by the commercial authorization of the first genetically engineered products in the early 1990s, the challenges attached to these new plant-breeding techniques are of a different nature, given that these technological developments are characterized by their quick pace and accessibility, the absence of foreign DNA sequences in the final products, and the impossibility of distinguishing these products from conventional ones using available detection methods (Lusser, Parisi, Plan, & Rodriguez-Cerezo, 2011).

As a result of these recent discoveries, various EU-level expert groups, advisory bodies, and agencies have provided their opinions on the novelty and the safety of these new plant-breeding techniques, aimed at providing some authoritative points of scientific reference. More concretely, the High Level Group of the Commission's Scientific Advice Mechanism (SAM) has published an independent Explanatory Note on "New Techniques in Agricultural Biotechnology."[2] Following a request from the College of Commissioners, led by Commissioner Andriukaitis for Health and Food Safety, the Group adopted, during their 5th meeting, the scoping paper on the topic "New techniques in agricultural biotechnology." The final piece of advice was delivered to the College of Commissioners on April 28, 2017. The resulting Explanatory Note performed a comprehensive scientific comparison that covered the whole spectrum of breeding techniques used for agricultural applications (conventional breeding techniques, established genetic modification and new breeding techniques), according to a defined set of aspects.

The European Academies Science Advisory Council issued three reports on the topic (EASAC, 2013, 2015, 2017) asking for full transparency in disclosing the process used and reinstating the need for an EU-wide regulatory intervention that targets the specific agricultural trait/product rather than the technology by which it is produced. The Council is of the opinion that it is essential to continue the commitment to phased research to assess the efficacy and safety of gene drives before it can be decided whether they will be suitable for use. This research must include robust risk assessment and public engagement. The latest EASAC report came to the conclusions that "the trait and product, not the technology, in agriculture should be regulated, and the regulatory framework should be evidence-based."

Before that, the European Commission had obtained scientific advice on new breeding techniques mostly via the study on "New Plant Breeding Techniques: State-of-the-art and Prospects for Commercial Development" carried out by the Commission's Joint Research Centre (Lusser, et al., 2011). The study, published in 2011, investigated the degree of development and adoption by the commercial breeding sector of new plant-breeding techniques, discussed drivers and constraints for further developments, and evaluated the technical possibilities for detecting and identifying crops produced by new plant-breeding techniques. An Expert Group of Member States established a list of new plant-breeding techniques and evaluated them in the light of the existing legislation and of the most recent available scientific data.

The EFSA Panel on GMOs also published, in 2012, scientific opinions on three techniques, namely cisgenesis, intragenesis, and site-directed nucleases techniques, in terms of the risks that they might pose and the applicability of the existing EFSA guidance documents on GM plants for their risk assessment.[3] EFSA concluded that the existing guidelines for risk assessment applicable to GM plants were also appropriate for cisgenic and intragenic plants, and for the ZFN-3 technique. EFSA also considered the hazards associated with cisgenic plants to be similar to those linked to conventionally bred plants, but that novel hazards could be associated with intragenic and transgenic plants (EFSA European Food Safety Authority, 2012). All these breeding methods could, however, "produce variable frequencies and severities of unintended effects, the frequency of which cannot be predicted and needs to be assessed case by case." The European Commission´s three Scientific Committees – SCHER, SCENIHR, and SCCS – also adopted three opinions on synthetic biology, focusing on its scope and definition, risk assessment methodologies, safety aspects, and research priorities.[4]

The plurality of scientific accounts on the safety and novelty of new forms of gene editing has not yet triggered any particular policy action at the EU level, in spite of the commonality of their findings. The long-awaited Commission´s legal interpretation of the regulatory status of products generated by new plant-breeding techniques has led to a wide-ranging institutional inertia and a revamping of the ethical debate about the permissibility of these scientific techniques.

WHAT IS ETHICALLY CHALLENGING ABOUT GENOME EDITING?

Beyond the pure, safety-driven scientific questions, gene editing, in its contemporary form, raises a series of ethical and regulatory questions. These include questions, such as whether and how gene editing should be used to make inheritable changes to the human genome, lead to designer babies, generate potentially risky genome edits or disrupt entire ecosystems (for more on this, see Caplan, Parent, Shen, & Plunkett, 2015; Cyranoski, 2015; Howard et al., 2013, Mulvihill et al., 2017; Somerville, 2015). As a result, the use of techniques such as CRISPR-Cas9 has led scientists to recommend a moratorium on making inheritable changes to the human genome as "an effective way to discourage human germline modification and raise public awareness of the difference between these two techniques" (Lanphier, Urnov, Haecker, Werner, & Smolenski, 2015, p. 411). For example, the application of CRISPR-Cas9 as a pest control technique may produce unintended effects and mutations, which may lead to the dispersion of gene drive, the disappearance of whole animal populations, accidental releases, and/or the irreversible disturbance of entire ecosystems. In fact, research activities, intended to modify the genetic heritage of human beings, which could make such changes inheritable, are not financed under Horizon 2020, the EU framework program for research and innovation.

Taking the non-maleficence principle into account in risk assessment, and distinguishing the clinical and therapeutic aims of gene editing from its enhancement applications/uses, has also become a major source of concern. Another important problem is the efficient and safe delivery of CRISPR-Cas9 into cell types or tissues that are hard to transfect and/or infect. Further, concerns include the prospect of irreversible harm to the health of future generations, and worries about opening the door to new forms of social inequality, discrimination, and conflict, as well as to a new era of eugenics. In relation to the latter, there are concerns that genome editing may be used to introduce, enhance, or eliminate traits for non-medical reasons potentially leading to what has been called "genetic classism" (Ledford, 2015).

Some of the ethical concerns caused by gene editing include issues that touch upon dignity, justice, equality, and discrimination,[5] proportionality and autonomy, sanctity of human life, and respect for human dignity, the moral status of the human embryo, and the potential crossing of the boundary to modifying the human germline (Cyranoski, 2015). Other concerns include the protection of vulnerable persons, the respect for cultural and biological diversity and pluralism, disability rights, the protection of future generations, equitable access to new technologies and health care, the potential reduction of human genetic variation, stakeholder roles and responsibilities in decision making, as well as how to conduct "globally responsible" science (German Ethics Council, 2017; Nuffield Council on Bioethics, 2016; UNESCO, 2015). Safety, dignity, proportionality, and justice, with regard to equity in the sharing of benefits and autonomy concerns, constitute the predominant ethical challenges across the entire spectrum of gene-editing applications.

These ethical challenges are further augmented by the sheer novelty introduced by genome-editing challenges. The affordability and user-friendly character of CRISPR-Cas9 make these innovative gene-editing techniques available to a large number of users, including many outside of institutional settings. The increased speed with which genome editing allows genetic manipulation to be achieved and the speed of its uptake and diffusion enhances its attraction. A series of ethical anxieties are centered upon the broad distribution, low cost, accelerated pace of development of this potentially dual-use technology, and intrinsic uncertainty about the downstream effects of gene editing that trigger the formation of a precautionary approach so as to tackle known hazards and unknown risks and prevent its deliberate or unintentional misuse (German Ethics Council, 2017; Hirsch, Lévy, & Chneiweiss, 2017; Mulvihill et al., 2017). Genome editing seems to distort the boundaries between basic, translational, and clinical research as well as the boundaries between therapeutic and non-therapeutic purposes. At the same time, it raises concerns "about possible misuse and abuses, in particular the intentional modification of human genome so as to produce individuals or groups endowed with particular characteristics and required qualities".[6]

Until now, apart from the European Group on Ethics (EGE; European Group of Ethics in Science and New Technologies EGE, 2015)), no other systemic initiative has been attempted at the EU level that could address the aforementioned ethical challenges from an institutional perspective in a comprehensive manner. The reasons behind this patchy approach include the general lack of EU competence on moral questions (Mohr, Busby, Hervey, & Dingwall, 2012). Several EU documents emphasize the competence of Member States as regards ethical issues and the variety of significant technical hurdles to be overcome before clinical or practical applications become a viable reality. Debating about technology ethics or even research ethics at the EU level has been seen as a continuous process of re-negotiation that ensures reciprocal non-interference between Member States as well as between Member States and the EU institutional order. Some of these open-ended questions on the morality and the ethical soundness of certain technological applications have been incorporated into specific pieces of EU Law such as the Clinical Trials Regulation (Regulation EU No 536/2014),[7] the EU GM-related Plant[8] and Food and Feed legislation,[9] the Regulation on Medical Devices,[10] and the Regulation on the EU Framework Programme on Research. In all these cases, where the ethical dimension is highlighted in the frame of the above mentioned pieces of EU secondary legislation, none of the *ad hoc* EU-level ethics committees/panels – formulated for example in the frame of the EU Framework Programme on Research– are allowed to provide any fully-fledged authorization or legally binding opinion. It is only national or local ethics committees that are empowered to act as the ultimate decision-makers or issue-frames on all matters of ethical gravity. The EGE issued a statement on Gene Editing[11] highlighting that there should be a moratorium on gene editing of human embryos or gametes that could result in the modification of the human genome. It acknowledged that the CRISPR-Cas9 system challenges the international regulatory landscape for the modification of human cells in the near to medium term.

Even if it is not an opinion *stricto sensu*, the statement of the EGE is of particular importance because it cautions against reducing the debate to safety issues and the potential health risks or benefits of gene-editing technologies. It highlights that many of the practical applications of gene editing will occur in the environmental sphere and will have significant implications for the biosphere. At the same time, it must be mentioned that despite some references to the opinions of the EGE in the context of the EU legislative process,[12] the influence of the work of this Group seems to be of marginal normative influence (Busby, Hervey, & Mohr, 2008; Plomer, 2008). This may be because it has no operational role in the realm of the systematic ethics reviews of proposals that raise ethical issues carried out in the Horizon 2020 framework, the Deliberate Release Directive, and/or the EU Clinical Trials context. In other words, neither EU Law nor the institutional advisory structures on ethics that have been established by the European Commission can exert a meaningful influence upon the way in which the various local or national narratives on the ethical soundness of gene editing are framed.

However, in view of the rapid pace of scientific developments in this field and the ongoing inter-institutional discussion on the main ethical tenets of the "successor" to the Horizon 2020 Framework Programme on Research, a normative point of reference and/or a certain form of ethical guidance on this rapidly evolving matter are needed. It would be irresponsible to proceed unless and until the relevant scientific, ethical, safety, and efficacy issues have been resolved and there are sufficient indications of a broad societal consensus. However, such a guidance, which may pre-determine the content or the interpretation of the relevant legal texts, should not generally be provided by transferring competences to committees composed of a small number of people or expert groups as these could be lacking in democratic legitimacy and accountability.

Given also the increasing number of bans and moratoria declared on gene-editing research across the world (UNESCO, 2015), the more pressing question for policy makers at present is whether germline genome editing technology research should be temporarily suspended, or under what conditions it could proceed. Varying views have been articulated. In general though, a fully-fledged prohibition will be nearly impossible to enforce due to the low cost of this novel technique and the plurality of ethical views on the topic.

REGULATORY QUESTIONS AND THE LEGAL STATUS OF GENE EDITING IN EUROPE

From a regulatory perspective, gene editing also raises thought-provoking questions about the adequacy of the current GMO framework to deal with all the challenges and questions that the latest scientific developments may bring forward and whether there is a need for regulation beyond the scope of existing GM regulations and other relevant areas of legislation. Are the current EU legal and risk analysis structures suitable to accommodate all of the concerns associated with the emergence of CRISPR-Cas9? Should safety or the foundational patent rights to CRISPR-Cas9 gene-editing technology (Contreras & Sherkow, 2017, p. 698; for more see, EPSO, 2015; Peng, 2016; Webber, 2014) be the main regulatory trigger for subjecting organisms to regulatory oversight? Do gene-edited organisms present a hazard or risk in a way that would necessitate regulatory oversight?

Some of the newest plant-breeding techniques are in an uncertain situation concerning their classification within the existing GM-related legislation (Abbott, 2015). There is considerable debate as to how these new techniques should be regulated and whether some or all of them should fall within the scope of EU legislation on GMOs given their potential to develop off-target events with still unknown consequences. It needs to be mentioned that, traditionally, European regulatory agencies pay particular attention to the method/scientific process that has been used. The result is that the scientific, ethical, and regulatory focus of all European stakeholders is on the *sui generis* features of their methods and techniques under development rather than on the relevant products.

Under EU law, the definition of GMOs states that "genetically modified organism (GMO) means an organism, with the exception of human beings, in which the genetic material has been altered in a way that does not occur naturally by mating and/or natural recombination" – Directive 2001/18/EC, Article 2(2). The annexes to the Directive further define the techniques that (1) result in genetic modification (listed in Annex I A, Part 1); (2) are not considered to result in genetic modification (Annex I A, Part 2); and (3) result in genetic modification but yield organisms that are excluded from the scope of the Directive (Article 3 and Annex I B): these techniques are mutagenesis and cell fusion (of plant cells of organisms that can exchange genetic material through traditional breeding methods). The 2001/18 EU Directive that governs the deliberate release of GMOs into the environment was drafted long before the latest gene-editing methods were developed and was based on clear differentiation between transgenic plants and conventional breeding.

The legal confusion arose because the 2001 Directive exempts organisms whose genomes have been altered by plant and animal breeders using mutagenesis techniques that were available at the time, such as irradiation. Therefore, the question arises if these Directives are suitable to face the new challenge of genetic engineering or if there is a need for updated regulations. Although genome-editing technologies facilitate efficient plant breeding without introducing a transgene, their applications seem to create indistinct boundaries in the regulation of genetically modified organisms given that the Directive 2001/18/EC contains both process- and product-related terms, it is commonly interpreted as a strictly process-based legislation.

The current debates over definitions and whether plants and nonhuman animals in which gene editing is performed are considered (legally) genetically modified organisms (GMOs) are particularly important to consider. A decision on whether genome-edited plants should be considered as GMOs has been pending in the European Union for some years. The Member States have asked the European Commission to issue guidance on the regulatory status of products generated using the new techniques, including whether gene-edited organisms should be considered, and regulated, as genetically modified organisms. A legal guidance on how to define plants produced by novel genome editing techniques is urgently needed as it may help genetically engineered organisms to escape from their current legal limbo in Europe. If the new techniques were to be exempted from GMO legislation, they would then also be exempt from the obligations of pre-market assessment and authorization, and from GMO labeling requirements. In Europe, only Sweden and the Netherlands have, so far, stated that CRISPR gene-edited crops or new plant-breeding techniques should not come under the scope of the existing EU legislative framework on GMOs. The European Parliament in its Resolution of June 7, 2016, on technological solutions for sustainable agriculture in the EU[13] considered it timely for the Commission to use its scientific findings as a basis for, *inter alia*, clarifying the legal status of the breeding techniques currently under scrutiny and to encourage open and transparent dialogu among all stakeholders on the responsible development of high-precision, innovative solutions for breeding programs.

There is also a case of significant importance for the inclusion or not of the novel gene-editing techniques under the scope of the current EU GMO framework that is pending before the Court of Justice of the European Union.[14] A request for a preliminary ruling was filed by the Conseil d'État (Council of State, France) in a case that involves the Confédération Paysanne: a French agricultural union defending the interests of small-scale farming as the main applicant and joined by eight other associations. Their objective is the protection of the environment and/or the dissemination of information concerning the dangers relating to GMOs. The Court has been asked to clarify the exact scope of the GMO Directive, more specifically the ambit, rationale, and effects of the mutagenesis exemption, and potentially assess its validity.

More broadly, the Court is invited to ponder the question of time: more precisely, the role that the passing of time and the evolution of technical and scientific knowledge should play with regard to both legal interpretation and the assessment of the validity of EU legislation. The Court is asked to rule on whether the exceptions for not applying the 2001/18/EC European Directive on the deliberate release into the environment of genetically modified organisms should be extended to the newest and more precise methods of targeted mutagenesis, including gene-editing techniques by ZFNs, TALENs, or CRISPR approaches. The ruling is also expected to shed light on whether Directive 2001/18 precludes Member States from adopting measures governing mutagenesis, including precautionary ones.

In addition to these ongoing reflections of a legal nature, it also needs to be mentioned that the European Medicines Agency (EMA) recently issued a concept paper on the revision of the Guideline on quality, non-clinical, and clinical aspects of medicinal products containing genetically modified cells.[15] It also published a draft guideline on safety and efficacy follow-up and risk management of Advanced Therapy Medicinal Products.[16] The EMA acknowledged that the introduction of the CRISPR-Cas9 system has rapidly increased the use of genome-editing technologies to genetically modify cells *ex vivo* for clinical applications, and it aims to take these aspects into consideration in its revised draft guideline, which is expected soon. It aims to reflect significant developments and experience gained since the publication of the current guideline. It intends to reassess the validity of the existing guidance text in the light of the current experience. It hopes to provide, where needed, specific quality, non-clinical, and clinical guidance for the development of CAR-73 T cells and related products. It will include considerations on the genome-editing tools when applied for the *ex vivo* genetic modification of cells.

Patenting CRISPR-Cas9 for therapeutic use in humans is also legally controversial. In February 2017, the US Patent and Trademark Office (USPTO) issued a decision on who should hold the patent on using CRISPR-Cas9 to edit genes, defining the terms and conditions for profit generation from this technology in future years. The risks of hereditary, unpredictable genetic mutations raise questions regarding the safety of the technique and the attribution of liability in case of damages. In a recent report, the US National Academies of Sciences, Engineering and Medicine urged caution when releasing gene drives into the open environment and suggested "phased testing," including special safeguards, given the high scientific uncertainties and potential ecological risks. Safety measures are necessary to avoid dissemination of organisms that may cause ecological damage or affect human health (National Academies of Sciences, Engineering, and Medicine, 2016).

In fact, many scientists caution that there is much to do before CRISPR could be deployed safely and efficiently. In particular, CRISPR might pose additional challenges from a risk assessment standpoint, in that organisms produced by these methods may create more pervasive changes to the genomes of living organisms than traditional genetic modification techniques.

In other words, in view of the upcoming ruling and legal interpretation, no particular regulatory or legislative action has been designed that could assess the suitability of the existing legal and regulatory framework at the EU level to correspond to the challenges of genome-editing applications. A regulatory fitness check might pave the way to the identification of the growing gap between these emerging technologies and the respective ethical and legal oversight structures. Until then, Member States will continue to identify legal ways for the permissibility or ban of all relevant research-related and commercial actions in this intriguing domain of the application of genetic engineering.

GENE DRIVE: GENE EDITING IN MOTION

An additional use of genome editing, particularly of the CRISPR-Cas9 tool, is envisioned in applications called synthetic "gene drives." As a general term, gene drive refers to DNA sequences that increase the frequency of their own inheritance and "has been rapidly adopted by the research community as a routine method to knock-in and knock-out DNA sequences in animals and plants" (Jones, 2015, p. 231). There are concerns that the spread of the gene drive will be difficult to control, and it might spread to populations or have consequences beyond those intended and these might not necessarily be beneficial. Gene drive has been associated with the production of variable frequencies and severities of unintended effects, the frequency of which cannot be predicted and needs to be assessed on a case-by-case basis (Resnik, 2014).

Thus, there is a need for emphasis on the importance of a phased approach to research to allow sufficient time to evaluate the efficacy and safety of gene drives before regulatory decisions can be made on whether or not they will be suitable for use. This phased research must include robust risk assessment and public engagement. This must involve those countries where gene drive systems would most likely be applied to tackle disease vectors. The risk assessment may need to focus on the following environmental concerns: potentially harmful (cross-border) environmental impact of releasing an organism with a gene drive into the environment; potentially rapidly spreading, permanent (irreversible) changes upon entire populations; spread and persistence; potential to cause irreversible ecological change; scientific uncertainties; possibility of horizontal transfer of the transgene(s) to non-target organisms; greater concerns over mobile genetic elements compared to "sterile" vectors and unforeseen consequences on human health.

Gene drives do not fit well within the existing EU regulatory rules on confinement and containment (European Parliament and European Council, 2009) as gene drives are designed to spread a genotype through a population. This makes confinement and containment much more difficult (or even irrelevant), as the environmental changes introduced by release are potentially irreversible. There is also a risk of possible dual use/malicious use given that a gene drive can also be used maliciously to cause the rapid spread in a population of a genetic product with a harmful effect. There is the potential to cause profound societal disruptions based on misuse, accident, or uncertain risks. As it is relatively simple to bring about a targeted genetic modification using CRISPR-Cas9 elements, there is a high potential risk for the inexpert use of these elements. Gene drives do not use proscribed agents or create regulated toxins, and thus fall beyond the scope of EU rules on dual use.[17]

Many regulatory questions arise in relation to gene drive. These relate to issues of subsidiarity, proportionality, precaution, sustainability, intergenerational equity, the regulation of both the research/development and the commercial stages, the trait, the main/reversal drive, the technology or the function, hazard, risk and uncertainty, and on which scientific grounds. Shall the regulatory intervention take place through codes of conduct or via a top-down regulatory instrument, liability clauses, and accountability provisions? It needs to be emphasized that there is little EU-wide agreement on a range of issues, including how risks should be framed, which methodologies should be adopted, and which values prioritized. There are also difficulties in accommodating socio-ethical concerns or non-positivistic accounts of science/expertise at the EU level, given that the nature of gene drives "raises many ethical questions and presents a challenge for existing governance paradigms to identify and assess environmental and public health risks" (National Academies of Sciences, Engineering, & Medicine, 2017, p. 7).

Within this frame, there is a need to consider the introduction of precautionary measures (moratorium, zero-risk approach, socio-ethical decision-making grounds, and technology acceptability) to perform basic and applied research on gene drives in a contained setting or in laboratories located in areas where the target species cannot survive or find mates. The decision on whether or not to utilize a gene drive for a given purpose should be based entirely on the probable benefits and risks of that specific drive. Building and testing a reversal drive for every primary drive that could spread a trait through a wild population: (1) improving data gathering and sharing in the face of limited resources; (2) filling newly exposed or created regulatory gaps; (3) incentivizing corporate responsibility and the introduction of self-regulatory schemes; (4) enhancing agency expertise and coordination; (5) providing for regulatory adaptability and flexibility; (6) achieving substantial, diverse stakeholder involvement (Chneiweiss et al., 2017; Mandel, 2009).

Among the main recommendations that have been framed for tackling the challenges of gene drive include the need to operationalize contextualized ecological risk assessments and to introduce a more dynamic/cyclical risk analysis model that is based on inclusive risk management and risk-benefit assessment in the context of the entire life cycle, from research to tangible utility/threat. A variety of diverse stakeholders – scientific, clinical, public health and military/security communities – should be involved in the drafting of community-based consent instruments that would approach technology as a value framework. A localized licensing approach should also be taken, which would echo developments in the international legal arena in the frame of the Cartagena Protocol, the Nagoya Protocol, and the various initiatives of the World Health Organization.

IS THERE A NEED FOR AN EU-WIDE REGULATORY INTERVENTION?

Given the ongoing discussion of the ethical tenets of gene editing, a question arises about whether there is a need for an EU-wide "intervention" that will address the socio-ethical challenges of techniques that could "expand the scope of genome editing" (Maeder & Gersbach, 2016, p. 446). First of all, when considering the design of a governance framework on gene editing, there is a need to take into account the historical experience in regulating such technologies and consider previous attempts to deal with genetic technologies. These include the multiple attempts to set up an operational framework on gene editing and the potential eugenic, dual use/misuse tendencies related to these technologies and the various conceptualizations of risk and/or uncertainty in the frame of this dynamic technological domain. The lack of a common trigger for regulating the products of genome editing and the absence of guidance regarding the data package required for risk assessment may stifle innovation and undermine "consumer confidence in both the risk assessment process and the safety of the biotechnology products" (Jones, 2015, p. 223).

At the same time, the current discussion is too focused on whether gene editing leads to organisms that are covered by the GMO definition in the first place and on the legal status of gene-edited organisms under current regulatory frameworks (European Parliament and European Council, 2001, 2003) without focusing on the moral dimension of modern genome editing. The *sui generis* features of this emerging technology trigger the need for the design of a governance framework that will approach gene editing as an object of regulatory action that does not have a scientific dimension *stricto sensu*. In other words, "the advances in genome-editing technologies mean that long-standing ethical questions can no longer be dodged on the basis of obvious and agreed safety concerns" (Mathews et al., 2015, p. 161). The debate should also extend to other, not necessarily science-based, risks and anxieties as it is crucial to establish regulatory standards that are centered "on the universally accepted principles enshrined in the UDBHR: human dignity; autonomy and individual responsibility; respect for vulnerable people and personal integrity; privacy and confidentiality; equality, justice and equity; non-discrimination and non-stigmatisation; respect for cultural diversity and pluralism; solidarity and cooperation; social responsibility for health; sharing of benefits; protection of future generations; protection of the environment, the biosphere and biodiversity" (UNESCO, 2015, p. 27).

Such a debate can contribute to the shaping of agricultural innovation in an inclusive and responsible manner and to seek a common point of equilibrium given that "public trust in science ultimately begins with and requires ongoing transparency, in terms of meaningful public input into the policy-making process related to human genome editing" (UNESCO, 2015, p. 27) and open discussion. In view of the growing "complexity" of scientific knowledge, the rather limited understanding of the concrete risks for health or the environment in the case of new breeding techniques, and the development of "diverse rationalities," which seem not to communicate with each other, the EU appears to have resorted to a proceduralized mode of governance expecting that this may lead to the formulation of a common governing narrative.

This proceduralized *modus operandi* is primarily focused on the establishment of a series of procedural modalities that facilitate the formulation of legal, ethical and scientific opinions that do not necessarily communicate with each other at the EU level. In the case of gene editing, this proceduralized approach has taken the form of the mobilization of the main EU-wide evidence-based procedural mechanisms. The growing accumulation of expert views on the risks, benefits and permissibility of genome editing in combination with the lack of a formal EU position or guidance on how gene editing should be approached illustrate the EU's structural limitations in framing an operational narrative on matters that are morally debatable. At the same time, it exhibits the limits of proceduralism as a tool that could accommodate value pluralism (Kritikos, 2017) but also the need for "the creation of a dynamic new political arena – in which reasoned scepticism is as valued in public debates about technology as it is in science itself" (Stirling, 2012).

Given the severe difficulties in regulating genetic engineering at the EU level during the last 30 years and the multiplicity of domains where gene editing could become applicable, any EU political initiative in this field should take into account the ethical tensions and significant regulatory concerns expressed by scientific unions, international organizations, and other stakeholders. It also needs to depart from a stringent science-based approach that has traditionally undermined the efficiency and legitimacy of its own interference on matters that are inherently political and social in their framing. An EU-wide initiative that will set the grounds for a responsible gene-editing narrative "could do away with polarizing product-versus-process and science-versus-values framings, and help to establish a governance system that is both informed by the science and guided by the concerns and values of citizens" (Kuzma, 2016). Such a system should be based on the need to re-enforce public trust and confidence in regulatory regimes in the domain of emerging technologies.

NOTES

1. See Directive (EU) 2015/412 of the European Parliament and of the Council of 11 March 2015 amending Directive 2001/18/EC as regards the possibility for the Member States to restrict or prohibit the cultivation of genetically modified organisms (GMOs) in their territory Text with EEA relevance, in force, OJ L 68, 13.3.2015, p. 1–8
2. Retrieved from http://ec.europa.eu/research/index.cfm?pg=newsalert&year=2017&na=na-280417
3. EFSA Panel on Genetically modified organisms (GMO); Scientific opinion addressing the safety assessment of plants developed using Zinc Finger Nuclease 3 and other Site-Directed Nucleases with similar function. *EFSA Journal* 2012; *10*(10): 2943. [31 pp.] doi:10.2903/j.efsa.2012.2943. Retrieved from www.efsa.europa.eu/efsajournal
4. SCENIHR (Scientific Committee on Emerging and Newly Identified Health Risks), SCHER (Scientific Committee on Health and Environmental Risks), SCENIHR (Scientific Committee on Emerging and Newly Identified Health Risks), SCCS (Scientific Committee on Consumer Safety), Synthetic Biology III – Research priorities, Opinion, December 2015; SCENIHR (Scientific Committee on Emerging and Newly Identified Health Risks), SCHER (Scientific Committee on Health and Environmental Risks), SCCS (Scientific Committee on Consumer Safety), Synthetic Biology II – Risk assessment methodologies and safety aspects, Opinion, May 2015;
5. US National Academy of Sciences, US National Academy of Medicine, Chinese Academy of Sciences, UK Royal Society. On human gene editing: International Summit Statement. Washington, DC, 2015.
6. Council of Europe. Statement by the intergovernmental Committee on Bioethics (DH-BIO) on gene editing technologies (2015) at 2.
7. Regulation (EU) No 536/2014 of the European Parliament and of the Council of 16 April 2014 on clinical trials on medicinal products for human use, and repealing Directive 2001/20/EC Text with EEA relevance, OJ L 158, 27.5.2014, p. 1–76.
8. Directive 2001/18/EC of the European Parliament and of the Council of 12 March 2001 on the deliberate release into the environment of genetically modified organisms and repealing Council Directive 90/220/EEC – Commission Declaration, OJ L 106, 17.4.2001, pp. 1–39.
9. Regulation (EC) No 1829/2003 of the European Parliament and of the Council of 22 September 2003 on genetically modified food and feed (Text with EEA relevance), OJ L 268, 18.10.2003 and Regulation (EC) No 1830/2003 of the European Parliament and of the Council of 22 September 2003 concerning the traceability and labeling of genetically modified organisms and the traceability of food and feed products produced from genetically modified organisms and amending Directive 2001/18/EC, OJ L 268, 18.10.2003, pp. 24–28.
10. Regulation (EU) 2017/745 of the European Parliament and of the Council of 5 April 2017 on medical devices, amending Directive 2001/83/EC, Regulation (EC)

No 178/2002 and Regulation (EC) No 1223/2009 and repealing Council Directives 90/385/EEC and 93/42/EEC (Text with EEA relevance.), OJ L 117, 5.5.2017, pp. 1–175.

11. Retrieved from https://ec.europa.eu/research/ege/pdf/gene_editing_ege_statement.pdf#view=fit&pagemode=none

12. 19th recital Directive 98/44/EC (note 31); 6th recital Commission Recommendation 2008/345/EC (note 9); 28th recital Regulation 1394/2007/EC on Advanced Therapy Medicinal Products [etc.], O.J. 2007, L 324/121, as amended by O.J. 2010, L 348/1; 33rd recital Directive 2004/23/EC on Setting Standards of Quality and Safety for the Donation, Procurement, Testing, Processing, Preservation, Storage and Distribution of Human Tissues and Cells, O.J. 2004, L 102/48, as amended by O.J. 2009, L 188/14.

13. European Parliament Resolution of 7 June 2016 on technological solutions for sustainable agriculture in the EU (2015/2225(INI). Retrieved from http://www.europarl.europa.eu/sides/getDoc.do?type=TA&language=EN&reference=P8-TA-2016-0251. Accessed on November 11, 2016.

14. Request for a preliminary ruling from the Conseil d'État (France) lodged on 17 October 2016 – Confédération paysanne, Réseau Semences Paysannes, Les Amis de la Terre France, Collectif vigilance OGM et Pesticides 16, Vigilance OG2M, CSFV 49, OGM : dangers, Vigilance OGM 33, Fédération Nature et Progrès v Premier ministre, Ministre de l'agriculture, de l'agroalimentaire et de la forêt, (Case C-528/16).

15. Retrieved from http://www.ema.europa.eu/docs/en_GB/document_library/Scientific_guideline/2017/07/WC500231995.pdf

16. Retrieved from http://www.ema.europa.eu/docs/en_GB/document_library/Scientific_guideline/2018/02/WC500242959.pdf

17. Council Regulation (EC) No 428/2009 of 5 May 2009 setting up a Community regime for the control of exports, transfer, brokering and transit of dual-use items, OJ L 134, 29.5.2009, pp. 1–269.

REFERENCES

Abbott, A. (2015). Europe's genetically edited plants stuck in legal limbo. *Nature*, *528*, 319–320. doi:10.1038/528319a

Baltimore, D., Berg, P., Botchan, M., Carroll, D., Charo, R. A., Church, G., … Yamamoto, K. R. (2015). Biotechnology. A prudent path forward for genomic engineering and germline gene modification. *Science*, *348*(6230), 36–38.

Bosley, K. S., Botchan, M., Bredenoord, A. L., Carroll, D., Charo, R. A., Charpentier, E., Cohen, R., Corn, J., Doudna, J., Feng, G., Greely, H. T., Isasi, R., Ji, W., Kim, J. S., Knoppers, B., Lanphier, E., Li, J., Lovell-Badge, R., Martin, G. S., Moreno, J., Naldini, L., Pera, M., Perry, A. C., Venter, J. C., Zhang, F., & Zhou, Q. (2015). CRISPR germline engineering – the community speaks. *Nature Biotechnology*, *33*(5), 478–486.

Busby, H., Hervey, T., & Mohr, A. (2008). Ethical EU law?: The influence of the European group on ethics in science and new technologies. *European Law Review*, *33*(6), 803–842.

Caplan, A. L., Parent, B., Shen, M., & Plunkett, C. (2015). No time to waste – The ethical challenges created by CRISPR. *EMBO Reports*, *16*, 1421–1426.

Chneiweiss, H., Hirsch, F., Montoliu, L., Müller, A. M., Fenet, S., Abecassis, M., … Saint-Raymond, A. (2017). Fostering responsible research with genome editing technologies: a European perspective. *Transgenic Research*, *26*(5), 709–713.

Contreras, J. L., & Sherkow, J. S. (2017). CRISPR, surrogate licensing, and scientific discovery. *Science*, *355*, 698–700.

Council of Europe. (2015). Statement by the intergovernmental Committee on Bioethics (DH-BIO) on gene editing technologies. In *DH-BIO/INF 13, 8th meeting*, Strasbourg, December 2, 2015, pp. 1–2. Retrieved from https://rm.coe.int/168049034a

Cyranoski, D. (2015). Ethics of embryo editing divides scientists. *Nature*, Mar 19; *519*(7543), 272.

EASAC. (2015). Statement on new breeding techniques. Retrieved from https://easac.eu/fileadmin/PDF_s/reports_statements/Easac_14_NBT.pdf

EASAC. (2017). EASAC report on Genome Editing: Scientific opportunities, public interests, and policy options in the EU. EASAC policy report 31, March 2017, ISBN: 978-3-8047-3727-3.

EASAC (European Academies Science Advisory Council). (2013). Planting the future: opportunities and challenges for using crop genetic improvement technologies for sustainable agriculture. European Academies Science Advisory Council Policy-Report 21, ISBN: 978-3-8047-3181-3. Retrieved from http://www.easac.eu/home/reports-and-statements/detail-view/article/planting-the.html

EFSA (European Food Safety Authority). (2012). Scientific opinion addressing the safety assessment of plants developed using zinc finger nuclease 3 and other site-directed nucleases with similar function. *EFSA Journal, 10*(10), 2943.

EPSO (European Plant Science Organization). (2015). Statement on crop genetic improvement technologies. Brussels, 26.2.2015, updated 18.12.2015. Retrieved from http://www.epsoweb.org/file/2147

European Group of Ethics in Science and New Technologies (EGE). (2015). EGE Statement on Gene Editing. Retrieved from https://ec.europa.eu/research/ege/pdf/gene_editing_ege_statement.pdf#view=fit&pagemode=none

European Parliament and European Council. (2001). Directive 2001/18/EC of the European Parliament and of the Council of 12 March 2001 on the deliberate release into the environment of genetically modified organisms and repealing Council Directive 90/220/EEC—Commission Declaration. *Official Journal of the European Union, 106*, 1–39.

European Parliament and European Council. (2003). Regulation (EC) No 1829/2003 of the European Parliament and of the Council of 22 September 2003 on genetically modified food and feed (Text with EEA relevance). *Off J L, 268*, 1–23.

European Parliament and European Council. (2009). Directive 2009/41/EC of the European Parliament and of the Council of 6 May 2009 on the contained use of genetically modified micro-organisms. *Official Journal of the European Union, 125*, 75–97.

German Ethics Council. (2017). Germline intervention in the human embryo: German Ethics Council calls for global political debate and international regulation. Retrieved from http://www.ethikrat.org/files/recommendation-germline-intervention-in-the-human-embryo.pdf

Hirsch, F., Lévy, Y., & Chneiweiss, H. (2017). CRISPR-Cas9: A European position on genome editing. *Nature*, Jan 4, *541*(7635), 30.

Howard, H. C., Swinnen, E., Douw, K., Vondeling, H., Cassiman, J. J., Cambon-Thomsen, A., & Borry, P. (2013). The ethical introduction of genome-based information and technologies into public health. *Public Health Genomics, 16*, 100–109. doi:10.1159/000346474

Jones, H. D. (2015). Future of breeding by genome editing is in the hands of regulators. *GM Crops & Food-Biotechnology in Agriculture and the Food Chain, 6*, 223–232. Retrieved from https://doi.org/10.1080/21645698.2015.1134405

Kaiser, J. (2017). U.S. panel gives yellow light to human embryo editing. *Science Health Policy*. Retrieved from http://www.sciencemag.org/news/2017/02/us-panel-gives-yellow-light-human-embryo-editing

Kritikos, M. (2017). *EU Policy-Making on GMOs: The False Promise of Proceduralism.* London: Palgrave-MacMillan.

Kuzma, J. (2016). Reboot the debate on genetic engineering. *Nature, 531*, 165–167. Retrieved from https://www.nature.com/news/policy-reboot-the-debate-on-genetic-engineering-1.19506

Lanphier, E., Urnov, F., Haecker, S. E., Werner, M., & Smolenski, J. (2015). Don't edit the human germ line. *Nature, 519*, 410–411.

Ledford, H. (2015). Where in the world could the first CRISPR baby be born? *Nature, 526*(7573), 310–311. Retrieved from https://www.scientificamerican.com/article/where-could-the-first-crispr-baby-be-born/

Lusser, M., Parisi, C., Plan, D., & Rodriguez-Cerezo, E. (2011). New plant breeding techniques: state-of-the-art and prospects for commercial development. *Joint Research Centre Technical Report EUR 24760*. Brussels: European Commission Joint Research Centre.

Maeder, M. L., & Gersbach, C. A. (2016). Genome-editing technologies for gene and cell therapy. *Molecular Therapy, 24*(3), 430–446.

Mandel, G. N. (2009). Regulating emerging technologies. *Law, Innovation and Technology, 1*(1), 75–92.

Mathews, D. J., Chan, S., Donovan, P. J., Douglas, T., Gyngell, C., ... Regenberg, A. (2015). CRISPR: A path through the thicket. *Nature, 527*(7577), 159–161.

Mohr, A., Busby, H., Hervey, T., & Dingwall, R. (2012). Mapping the role of official bioethics advice in the governance of biotechnologies in the EU: The European Group on Ethics' opinion on commercial cord blood banking. *Science and Public Policy, 39*(1), 105–117.

Mulvihill, J. J., Capps, B., Joly, Y., Lysaght, T., Zwart, H. A. E., & Chadwick, R. (2017). International Human Genome Organisation (HUGO) Committee of Ethics, Law, and Society (CELS). Ethical issues of CRISPR technology and gene editing through the lens of solidarity. *British Medical Bulletin, 122*(1), 17–29.

National Academies of Sciences, Engineering, and Medicine. (2016). *Gene drives on the horizon: advancing science, navigating uncertainty, and aligning research with public values.* Washington, DC: The National Academies Press. Retrieved from https://doi.org/10.17226/23405

National Academies of Sciences, Engineering, and Medicine. (2017). *Human Genome Editing: Science, Ethics, and Governance.* Washington, DC: The National Academies Press, https://doi.org/10.17226/24623

Nuffield Council on Bioethics. (2016). Genome editing: An ethical review. Retrieved from http://nuffieldbioethics.org/wp-content/uploads/Genome-editing-an-ethical-review.pdf

Peng, Y. (2016). The morality and ethics governing CRISPR-Cas9 patents in China. *Nature Biotechnology, 34*(6), 616–618.

Plomer, A. (2008). The European Group on ethics: Law, politics and the limits of moral integration in Europe. *European Law Journal, 14*, 839–859.

Reardon, S. (2016). First CRISPR clinical trial gets green light from US panel. *Nature News*, June 22, 2016.

Resnik, D. B. (2014). Ethics of community engagement in field trials of genetically modified mosquitoes. *Developing World Bioethics, 14*(1), 37–46.

Somerville, M. (2015). Debating the ethics of gene editing; in: Brown J (ed.): *The 180*, 2015.

Stirling, A. (2012). Opening up the politics of knowledge and power in bioscience. *PLOS Biology, 10*(1), e1001233.

The Hinxton Group. (2015). Statement on genome editing technologies and human germline genetic modification. Retrieved from http://www.hinxtongroup.org/Hinxton2015_Statement.pdf

UNESCO. (2015). Report of the International Bioethics Committee on updating its reflection on the human genome and human rights. SHS/YES/IBC-22/15/2 REV.2 Paris, 2 October 2015. Retrieved from http://unesdoc.unesco.org/images/0023/002332/233258E.pdf

US National Academy of Sciences. (2015). US National Academy of Medicine, Chinese Academy of Sciences, UK Royal Society. On human gene editing: International Summit Statement. Washington, DC. Retrieved from http://www8.nationalacademies.org/onpinews/newsitem.aspx?RecordID=12032015a

Webber, P. (2014). Does CRISPR-Cas open new possibilities for patents or present a moral maze? *Nature Biotechnology, 32*(4), 331–333.

ARRIGE: TOWARD A RESPONSIBLE USE OF GENOME EDITING

François Hirsch and Lluis Montoliu

ABSTRACT

For more than 20 years, genome editing has been one of the numerous technologies developed for the study and manipulation of the genome. However, since the relatively recent appearance of the so-called precision approaches, and especially through the "CRISPR revolution," the modification of the genome of any living beings on our planet has become possible, despite recent results showing some unexpected and undesirable effects of this technology. The objective of this chapter is to illustrate how a mobilization of the scientific community through the setting-up of an association should allow a responsible and ethical use of these technologies with considerable impacts for our society.

Keywords: CRISPR-Cas; TALEN; ZFN; responsible research and innovation; livestock; gene therapy; crops

INTRODUCTION

In 1993, Francisco Martínez Mojica, a microbiologist of the University of Alicante (Spain), reported some unusual DNA repeated elements found in the genome of an archaea living in the salt marshes of Santa Pola, a village near Alicante (Mojica, Juez, & Rodríguez-Valera, 1993). Twelve years later, after analyzing many similar DNA structures found in several prokaryotes, Mojica and coworkers proposed these DNA arrays would be the basis of an adaptive immune defense used by bacteria and archaea to fight any intruder element, including bacteriophages and plasmids (Mojica, Díez-Villaseñor, García-Martínez, & Soria, 2005). He had previously coined a unique acronym to name

Ethics and Integrity in Health and Life Sciences Research
Advances in Research Ethics and Integrity, Volume 4, 115–127

ISSN: 2398-6018/doi:10.1108/S2398-601820180000004008

these peculiar DNA repeated arrays: CRISPR (Clustered Regularly Interspaced Short Palindromic Repeat) (Mojica & Montoliu, 2016). When bacteria are attacked by a virus, they react by cutting the viral DNA genome in at precise locations and preserve the memory of these encounters by keeping these fragments integrated in the bacterial genome, within the CRISPR array. Thus, this CRISPR locus behaves as a kind of hard disk for storing these fragments. Moreover, all these inserted small DNA pieces get routinely transcribed and can be found in the cytoplasm. Subsequently, when individuals derived from that original bacterium are again attacked by the same virus, it already has its "identity card." The transcribed fragments match their DNA sequences with that of the infectious agent. This recognition triggers the recruitment of a "molecular scissor," a RNA-guided DNA endonuclease, a Cas (CRISPR associated) enzyme, which eventually cuts out and destroys the intruding viral genome. The bacteria are thus protected from the virus by this powerful and adaptive immune system with a genetic basis. Eventually, the underlying mechanism of the CRISPR system was fully understood and their constituents characterized, allowing its conversion into an efficient genome-editing tool.

OTHER ADVANCES OF GENOME-EDITING TECHNIQUES

In parallel, other groups developed different gene-editing approaches based on the use of analogous systems, such as Zinc Finger Nucleases (ZFNs) and Transcription Activator-Like Effector Nucleases (TALENs); however, they remain quite expensive and tedious (Wood et al., 2011). However, the three systems (ZFN, TALEN, and CRISPR) are operating under similar mechanisms, with an endonuclease digesting DNA at specific locations, guided by other helper elements, namely artificial or natural proteins in the case ZFN and TALEN, respectively, or small RNA molecules, in the case of CRISPR tools. The ZFN and TALEN technologies, owing to the fact they were described well before the current CRISPR tools, some 10 years earlier, have even been already transferred to human clinical trials in various pathologies.[1]

CRISPR research progress silently continued within few laboratories up to 2012 when several teams demonstrated that this technique could be used for the target modification of the genome of eukaryotic cells, accurately, inexpensively and rapidly (Doudna & Charpentier, 2014). In just a few years, the CRISPR technique has almost substituted for the others, being applied to human, animal, fungal and plant cells, resulting into a large variety of applications.

Sheep may be edited to gain a double-muscle phenotype (Crispo et al., 2015). Genome-edited pigs can be protected against devastating viruses, which cost billions of euros to farmers (Whitworth et al., 2016). Convenient traits impacting on animal welfare, already found in existing races or varieties, can be effectively mobilized into productive strains, as was the case with the hornless dairy cattle obtained after reproducing the known *Polled* mutant allele through TALEN-mediated genome edition (Carlson et al., 2016). Fungal species may also be edited, for instance to improve mushroom conservation, inhibiting the spontaneous oxidation process upon one of the first genome-edited food products not regulated by the US Department of Agriculture.[2] The US company DowDuPont-Pioneer has decided to scale up the first plants genetically edited by the CRISPR technique.[3]

In humans, the first report came from the article published in August 2015 by a Chinese team demonstrating the possibility of manipulating in vitro fertilization (IVF)-derived and discarded triplonuclear human embryos using CRISPR (Liang et al., 2015). These researchers attempted to modify the gene responsible for β-thalassemia;[4] however, only a few of the manipulated embryos expressed the mutated gene. In addition, they found all the same concerns regarding potential on-target (genetic mosaicism) and off-targets that had been previously detected in other mammalian species, in embryos, mostly rodents. Despite the limitations of the proof-of-concept approach chosen by these researchers, and the subsequent publications that appear on the subject, mostly from China, the demonstration is made that human embryos can be modified in a simple way. In August 2017, an article published by several American teams led by Shoukrat Mitalipov, in collaboration with other researchers in Korea, followed reinforcing the idea that human embryos could be efficiently modified by CRISPR technology (Ma et al., 2017), although some interpretations of this publication have been challenged and robustly criticized by other researchers (Egli et al., 2017). The study by Mitalipov also represented the first time that human embryos had been created ad hoc for research purposes, an illegal procedure in most EU countries adhering to the Convention for the Protection of Human Rights and dignity of the human being with regard to the applications of biology and medicine, signed in Oviedo on April 4, 1997.

Regarding other uses of CRISPR in human embryos, 2016 marks the date of the first authorization given by a European regulatory agency, the UK Human Fertilisation and Embryology Authority (HFEA), for the manipulation of IVF-derived and donated human embryos in the context of research (Callaway, 2016). This first European permission, granted to Kathy Niakan's laboratory (The Crick Institute, London, UK) resulted in a very informative publication demonstrating different behavior of the same gene product (Oct4) in rodent and primate preimplantation embryo development (Fogarty et al., 2017). This experiment, and the unexpected findings, illustrates why research is needed in preimplantation human embryos, and the mistakes and misassumptions that could be caused by simply inferring human embryo development from the observed rodent embryo development, without experimental evidences. Similarly, the US National Institute of Health has agreed to sponsor the first clinical trial using CRISPR/Cas9-modified human cells in cancer,[5] following similar developments pioneered in China, aiming to inactivate the PD-1 locus, with the objective of reactivating the patient's immune response against tumor cells.

CRISPR technology has also impacted the biomedical research field, where it is now possible to reproduce mutations diagnosed in patients affected by any congenital disease into cellular or animal models. Examples of genome-edited laboratory (pre-clinical) models for investigating human diseases include albinism (Seruggia, Fernández, Cantero, Pelczar, & Montoliu, 2015), cancer (Huang et al., 2017), Duchenne muscular dystrophy (Nelson et al., 2016), metabolic liver diseases (Yang et al., 2016), retinal degenerative diseases such as retinitis pigmentosa (Suzuki et al., 2016), neurodegenerative diseases such as Huntington (Yang et al., 2017), severe skin disorders such as dystrophic epidermolysis bullosa (Hainzl et al., 2017), glaucoma (Jain et al., 2017), and many other pathologies. The ease by which many human mutant alleles have been replicated in cells or laboratory animals has been mainly due to the flexibility and robustness of the CRISPR genome-editing tools.

EXTENDED APPLICATIONS OF GENOME EDITION

All these pre-clinical attempts have been instrumental to illustrate the power of the CRISPR genome-editing techniques for innovative gene therapy protocols but have also highlighted the need to further basic research on the fundamental mechanisms involved in DNA repair, essential to understand and be able to control the ratio between expected/desired versus unexpected/undesired mutations. The fact is that most CRISPR genome-editing attempts trigger the Non-Homologous-End-Joining repairing path, which proceeds randomly, thereby inserting and deleting (INDELs) nucleotides and resulting in a variety of alleles generated.[19] This genetic multiplicity or inherent mosaicism is the main trouble that genome-editing techniques will have to fix before these approaches can be exported, safely and efficiently, for their use in human beings.

Besides the application aiming to address human pathologies, one additional field with potential applications of CRISPR technology is the eradication of animal species harmful to humans by the so-called "gene drive" approach. This is based on a Super-Mendelian inheritance driven by a gene construct encoding all CRISPR-Cas9 reagents and flanked by homologous sequences surrounding the target DNA aimed by the RNA guide. The system is prone to auto-perpetuation since each mutant allele is copied onto the homologous chromosome. In the best scenario, carriers of gene-drive mutant alleles only transmit mutant alleles at 100% through their progeny, rapidly extending the mutant allele among a population. The first CRISPR-based gene-drive protocol was tested in *Drosophila* (small fruit flies), maintained in a contained biosafety laboratory (Gantz & Bier, 2015).

The idea behind these proposals is that CRISPR would help ending the endemic insect-borne diseases that kill millions of people every year, those living mainly in the world's weakest economies. These include mainly malaria (Amouzou, Buj, Carvajal, Cibulskis, & Fergus, 2015), dengue, chikungunya, yellow fever, or Zika virus diseases. Eradication of anopheles, the disease vectors, has already occurred at different sites (McGraw & O'Neill, 2013). In Brazil, the originally British biotechnology company Oxitec, now owned by the US company Intrexon, proposes to release male anopheles modified by more conventional gene therapy techniques allowing the insertion of a killer gene that will affect the offspring and the females encountered. Thus, the use of the OX513A mosquito makes it possible to obtain interesting results on the dengue epidemic but at very high costs (Alphey et al., 2013). Similarly, in Africa and Burkina Faso, further trials are under way in the fight against malaria.[6]

In parallel, several research laboratories are developing an approach based on CRISPR technology, which would allow faster, more efficient, and cheaper results. The approaches chosen are the integration of gene modifications rendering the mosquitoes sterile with very interesting results but not mature enough for rapid transfer to life-size applications (Hammond et al., 2016). In fact, unanticipated aspects have been already been encountered in laboratory population boxes where some pilot experiments have been run, before releasing these genome-edited mosquitoes into the wild. These experiments indicate, invariably, that all gene-drive approaches are eventually displaced away, as soon as some mutant target alleles, incorrectly repaired, are no longer target for further modification, thus becoming resistant and again rapidly dispersed among the entire population (Hammond et al., 2017).

In addition to those approaches that might have a real impact on the public health of many developing countries, New Zealand has decided to launch a national "Predator-free 2050" plan for the eradication of animals considered as predators, demonstrating that humans are ready to use CRISPR technology to increase its dominance over Nature.[7] However, it might be too early to launch these gene-drive applications to eradicate mammals. Recent reports warn that gene-drive approached might function very differently in vertebrates as compared to invertebrates, to insects where these methods were first tested. It appears that the self-copying activity of the gene-driven mutant allele only occurs in a subset of individuals and only when the Cas9 endonuclease is expressed in the female germline, not the male one (Grunwald et al., 2018). These results would indicate that we are still far from having control on these techniques before they can be safely deployed in the nature.

MARKETABLE PRODUCTS DERIVED FROM GENE-EDITING DEVELOPMENTS

The most recent significant aspect of the "CRISPR revolution" is its industrialization, its commercial use by companies willing to benefit, legitimately, from their use in the market. This process now goes through a fight without mercy around intellectual property. Hardly before the first results were obtained on the undeniable potentialities of this technology, the world of biotechnologies started to prepare the commercialization of the products that will come from it. Jennifer Doudna, Emmanuelle Charpentier, and Feng Zhang, co-discoverers of CRISPR applications in mammals, have created their own start-ups (respectively, Caribou, CRISPR Therapeutics and Editas Biotechnologies), and their home institutions behind, mostly Berkelely University and BROAD Institute-MIT. In a few years, many patents or patent families have been filed, and the real battle has taken place in Europe and the United States, between law firms that attack their paternity.[8] Such effervescence observed so soon after the first publications and in such a sharp biological field had never been seen. Some fear that this battle is already delaying the progress of this field in the private sector, where the good laboratory ideas need to crystalize to become useful and practical applications.

Before addressing the societal-ethical aspects that such a scientific breakthrough can only provoke, it is important to mention several controversial works that seem to underline the limits of this technology. The first concerns the demonstration of unintentional changes (also known as off-target) whose abundance and lack of control could have represented and triggered serious consequences when using CRISPR tools for in vivo experiments (Schaefer et al., 2017). However, this article by Kellie et al., published in *Nature Methods*, was quickly questioned for its methodological bias and interpretation problems and ended up being retracted by the publisher (***Schaefer et al., 2018). In brief, the authors misinterpreted the information of genetic variants they collected. They assumed these were directly related to the effect of CRISPR-Cas9 tools on the mouse genome, but the reality was that they inadvertently documented genetic variants whose origin was the spontaneous mutations regularly accumulating on any organism and at every generation, mixed up with the genetic drift consequences of selecting individuals randomly at every generation. Therefore, those mutations were expected and unrelated to CRISPR-Cas9 technology (Montoliu & Whitelaw, 2018).

More serious concerns were reported by Matt Porteus and collaborators from Stanford University. They discovered the presence of T-lymphocytes and antibodies against Cas9 protein from *Streptococcus pyogenes* and *Staphylococcus aureus* (the two most common sources of endonuclease used world-wide in any genome-editing experiment), in the majority of human serum samples they tested, hence suggesting that our prior contact with these pathogens had already triggered the development of an immune reaction against them and their compounds (Charlesworth et al.). This finding recommends caution and careful reconsideration before using these Cas9 reagents for in vivo therapy in patients. Although Cas9 is not the only enzyme of this kind that can be used, this discovery does not call into question the CRISPR approach as a whole, but it has given rise to a panic in the stock market severely affecting the value of the biotechnology companies developing this approach. Proposed solutions for overcoming this problem are the combined use of immunosuppressive drugs or, better, the use of Cas-like nucleases from other prokaryotes without previous known interactions with humans, such as Cpf1, now called Cas12a (Zetsche et al., 2015).

More recently, two publications in the prestigious journal *Nature Medicine* of June 2018 have created a great stir within the scientific community and again in the stock markets (Haapaniemi, Botla, Persson, Schmierer, & Taipale., 2018; Ihry et al., 2018). These reports further documented a previous observation, that of the conflicting activities between Cas9-induced double-strand breaks (DSB) on DNA and p53-monitoring of these DSBs, and its immediate promotion for repairing. P53 is the guardian of the genome and its activity is required to promote repairing, senescence, or apoptosis of cells depending of the damaged detected. Of course CRISPR-Cas9 trigger some DNA damage, and while this activity is partly counteracted by endogenous P53, we simply have to learn how to deal with these two activities in order to find a good balance. We cannot promote the inhibition or deactivation of P53 gene because this would result in worse consequences, with a high risk of the cells developing a tumor behavior. All of these results and those undoubtedly to come, commit scientists to show some caution before embarking on the first in vivo trials in humans or animals.

RISKS AND REGULATION OF GENOME EDITING

One of the consequences not foreseen by the discoverers of CRISPR activities is its potential uncontrolled use by individuals, outside official laboratories, in "garage"-like settings, at home, or in countries without effective monitoring or surveillance. In 2016, the Director of the US intelligence services defined CRISPR technology as a potential mass destruction weapon.[9] Another uncontrolled usage of CRISPR is triggered by the international Do-It-Yourself (DIY) movement. This movement launched in North America in the late twentieth century promotes, in terms of technology, their provision for all. For example, a young Californian, Josiah Zayner, who defines himself as a "biohacker," filmed himself injecting a genetic construct prepared using a CRISPR commercial kit in order to enrich his muscle mass through the inactivation of the myostatin gene. He defined himself as the first human modified by genome editing.[10] Of course these actions are generally useless, not recommended, and can likely cause more damage (through infections and immunoreaction) than potential benefit. These cases suggest that access to CRISPR reagents must be regulated and only distributed to legitimate end-users, as with many other products for research and medical use, which cannot be obtained in the market unless purchased from an approved user to an approved vendor.

Several international bodies have begun to think about the societal and ethical aspects of the rapid development of this technology. The first aspect concerns the status of products processed by genome editing: should they be considered genetically modified organisms (GMOs) which would make their use very framed from a regulatory point of view, or should not be regulated and be considered "New Breeding Techniques" (NBT) which would escape these constraints?

From a scientific point of view there are no substantial differences from the mutations triggered by physical agents (radiation), which are not regulated, or by genome-editing tools like CRISPR. Furthermore, the end-product after a genome-editing experiment is an organism carrying a mutant allele similar, or sometimes identical, to others existing in nature. Therefore, it would not appear justified to apply different rules to very similar varieties that their main difference was to have been obtained through different methods. These contradictions were first reported by Stefan Jansson, from Umea University (Sweden), using some Arabidopsis and cabbage genome-edited plants (Jansson, 2018).

As discussed above, the USDA considered CRISPR-modified mushrooms to be non-GMO and decided not to regulate the use of these products.[11] In Europe, first Sweden and then The Netherlands have taken the decision not to regulate products from genome-editing technologies pending the final decision of the European Commission. We have just learnt the sentence of the Court of Justice of the European Union[12] seized by the French Council of State, and it states that genome-edited organisms should be regarded as GMO and regulated through the corresponding European Directive 2001/18/EC as any other transgenic organism. The implications of this sentence are yet to be analyzed, but the economic and societal stakes are considerable. NBT proponents had claimed that genome editing does not involve the insertion of exogenous genetic material as for GMOs, but the rewriting (or deletion) of certain nucleotides. Instead, the application of the precautionary principle has forced the considering of genome-editing organisms as GMO in the Europe Union, undertaking an opposite route as preliminary announced in the USA.

Whatever the regulatory status of these products, however, there are still major concerns about, as mentioned above, including the finally feasible irreversible modification of the human genome and the possibility of eradicating animal species. In 2015, representatives of several academies met in Washington DC to call for more international awareness on these issues while refusing to propose a moratorium that in the current conditions of the expansion of these techniques seemed unrealistic.[13] However, in the context of the 13th Conference of the Parties to the Convention on Biological Diversity in Cancún, Mexico, in December 2016, more than 170 nongovernmental organizations requested a moratorium on the modification of animal genomes for eradication genetic forcing, as long as the ecological impacts were not precisely appreciated.

CONCLUSION: TAKING ACTION

In Europe, many academies and institutions also raised the question of the merits of challenging the inviolability of the human genome as provided for in Article 13 of the Convention for the Protection of Human Rights and dignity of the human being with regard to the applications of biology and medicine, signed in Oviedo on April 4, 1997.[14] However, the technique is of great interest to associations of patients with severe life-threatening genetic diseases (Alzheimer's, Parkinson's and Huntington's disease, etc.) who want the rapid launch of clinical trials in humans.[15] More worryingly, it is also of interest to the proponents of transhumanism, who find in it the much-desired method of modifying heredity with, of course, all the potential improvement, enhancement, and eugenic excesses.

Everyone agrees that genome editing is a transformative technology. As discussed above, numerous applications have been already explored in cellular and animal models, and in plants and other species for biological and biotechnological purposes. However, biomedical applications, aiming to potentially treat and cure many genetic diseases, have yet to become a reality, because these novel therapeutic approaches require the careful evaluation of safety and efficacy parameters before being exported to the clinic, for use on patients.

In spite of the endless potential benefits of CRISPR-Cas genome-editing applications, farmers, patients, and many other citizen groups are largely unaware of the effects, risks, and profound implications associated with the heritable modification of genomes from any species, including the human genome. Common discussions appear to be associated with nonmedical usages of CRISPR techniques, whereas medical uses of new technologies usually encounter better acceptance in society.

The immediate past showed us some examples of communication schemes that proved wrong and counterproductive, such as the discussions around the benefits or risks associated with growing and eating transgenic plants. To avoid repeating these mistakes, the adoption of transparency initiatives is currently considered a better strategy. This has proven to be the case behind the change in societal perception of animal experimentation (Montoliu, 2018). Raising awareness, sharing information with clarity and openness, is the fundamental first step to allow public stakeholders to debate and judge the many uncertainties and transformative potential of genome-editing technologies.

In France in 2015, following a referral to its CEO, the INSERM (the French National Institute for Health and Medical Research) Ethics Committee (IEC) initiated a national and international debate, also involving experts living in economically vulnerable countries who are or will be impacted by these technologies.[16] With the moral and economic support of the INSERM IEC, this group of researchers and ethicists from France and neighboring countries organized and held a series of meetings in Europe, and overseas, in India, Africa, and South America, forming part of the so-called Global South, in order to address the ethical issues associated with the responsible use of genome-editing techniques, as applied to humans, animals, plants, and the environment. The group's initial position was first published in *Nature* in early 2017 (Hirsch, Lévy, & Chneiweiss, 2017). This initial proposal was further elaborated upon discussing these issues within a broader group of interested participants. Eventually these discussions crystalized into a second publication, in *Transgenic Research*, released in July 2017 (Chneiweiss et al., 2017), where the group proposed the creation of a European Steering Committee to assess the potential benefits and risks of genome-editing methods. Additional aims for this Committee would be designing risk matrices and scenarios for responsible uses of this technology, and contributing to an open debate on societal aspects prior to a translation into national and international legislation.

The following meeting of this group, promoted from INSERM, took place in Paris, in November 2017, to evaluate the many existing reports, position papers, and manifests on the same subject, on the ethical and societal aspects of genome editing, and the responsible uses of these technologies, as nicely summarized in this review by de Lecuona and collaborators (de Lecuona, Casado, Marfany, Lopez Baroni, & Escarrabill, 2017). At that meeting, the attendees expressed their majority will to go beyond the publication of a report, in order to become a useful and operative association, proactively engaging all different stakeholders within the genome-editing applications and uses scope. The participants decided to expand the group from Europe to a true international representation, aiming for an integrative approach that would include perspectives from Europe, North America, China, Japan, and Australia, as well as the rest of the world, often forgotten, and geographically located in the Global South, including members from Southeast Asia, Africa, and Central/South America. All these previous discussions and debates became a reality on March 23, 2018, in a new conference held at Île-de-France regional Parliament in Paris, where more than 160 participants gathered from 35 countries to openly discuss all these topics. One of the main decisions adopted by this meeting in Paris was to launch the ARRIGE (Association for Responsible Research and Innovation in Genome

Editing) initiative, as it was described in a collaborative article published in the second issue of *The CRISPR Journal* (Montoliu et al., 2018).

The ARRIGE initiative triggered an enormous interest among the scientific community and several articles featured and described this new association, in relation to other similar initiatives that were launched in the USA (Hurlbut et al., 2018; Saha et al., 2018), including reports published in *Science* (Enserink, 2018) and *Nature Biotechnology* (Smalley, 2018). Correspondingly, the ARRIGE Steering Committee created and launched a dynamic web site for the newborn association, along with registering the presence of the group in numerous social networks and through different communication channels (please see https://arrige.org for further information).

The aim of ARRIGE, as a new nonprofit initiative, is to promote a global governance of genome editing through a comprehensive setting for all stakeholders – academics, researchers, clinicians, public institutions, private companies, patient organizations, and other nongovernmental organizations, regulators, citizens, media, governmental agencies, and decision-makers from all continents. ARRIGE hopes to address multiple issues raised by genome-editing technologies used in research and applications within a safe and ethical framework for individuals and society.

More specifically, the ARRIGE association aims to provide a vehicle for meetings and outreach with the following four major objectives:

(1) fostering an inclusive debate with a risk-management approach, considering human, environmental, animal, and economic issues;
(2) getting involved in the governance of genome-editing technology with governmental and intergovernmental stakeholders;
(3) creating an ethical toolbox and informal guidance for genome-editing technology users, regulators, governance, and the civil society at large, including those living in low- and middle-income countries; and
(4) developing a robust and specific reflection on the role of the lay public in this debate and the necessity for improved public engagement.

In practical terms, the mission of ARRIGE is to inform the general public and decision-makers about the real issues involved in the development of genome-editing techniques, by creating thematic think tanks. This reflection will focus on the search for methods to evaluate the efficacy and safety of techniques, on the evaluation of the impacts of the use of "gene drive," on the acculturation of the scientific community to the development of responsible research, and fair communication to the general public.

The CRISPR wave has triggered a revolution and requires investing in awareness similar to that aroused by the first genetic manipulations in the 1970s, which resulted in a moratorium proposed by the greatest scientists at the famous Asilomar conference. Thankfully, it seems that all developers of CRISPR technologies have decided to engage in an open and responsible debate with society, and this is the arena where ARRIGE will deploy its activities. ARRIGE participants would show that science may be in society and not out of society. It is our duty and our responsibility.

NOTES

1. Retrieved from https://clinicaltrials.gov/ct2/results?cond=&term=genome+editing&cntry=&state=&city=&dist=&Search=Search
2. Retrieved from https://www.scientificamerican.com/article/gene-edited-crispr-mushroom-escapes-u-s-regulation/
3. DuPont Pioneer Establishes a CRISPR-Cas Advanced Breeding Platform. Retrieved from http://seedworld.com/dupont-pioneer-establishes-crispr-cas-advanced-breeding-platform/
4. β-thalassemia is a blood disorder that reduces the production of hemoglobin which leads to a lack of oxygen in many parts of the body.
5. Retrieved from http://www.nature.com/news/first-crispr-clinical-trial-gets-green-light-from-us-panel-1.20137
6. Retrieved from http://www.radarsburkina.net/index.php/societe/373-lacher-de-moustiques-genetiquement-modifies-au-burkina-faso-est-ce-la-panacee-pour-l-eradication-du-paludisme
7. New Zealand's War on Rats Could Change the World. Retrieved from https://www.theatlantic.com/science/archive/2017/11/new-zealand-predator-free-2050-rats-gene-drive-ruh-roh/546011/
8. How the battle lines over CRISPR were drawn. Science, Feb. 15, 2017. Retrieved from http://www.sciencemag.org/news/2017/02/how-battle-lines-over-crispr-were-drawn
9. Top U.S. Intelligence Official Calls Gene Editing a WMD Threat. MIT Technology Review. Retrieved from https://www.technologyreview.com/s/600774/top-us-intelligence-official-calls-gene-editing-a-wmd-threat/
10. Retrieved from http://www.ifyoudontknownowyaknow.com
11. USDA confirms it won't regulate CRISPR gene-edited plants like it does GMOs. Retrieved from https://newatlas.com/usda-will-not-regulate-crispr-gene-edited-plants/54061/
12. Retrieved from https://curia.europa.eu/jcms/upload/docs/application/pdf/2018-07/cp180111en.pdf
13. International summit on human gene editing: a global discussion; Retrieved from http://nationalacademies.org/gene-editing/Gene-Edit-Summit/
14. Retrieved from https://www.coe.int/en/web/conventions/full-list/-/conventions/treaty/164
15. Basic understanding of genome editing: The report. Retrieved from http://www.geneticalliance.org.uk/news-events/news/understanding-genome-editing/
16. Retrieved from https://www.inserm.fr/recherche-inserm/ethique/comite-ethique-inserm-cei/notes-comite-ethique-en-reponse-saisines

REFERENCES

Alphey, L., McKemey, A., Nimmo, D., Oviedo, M. N., Lacroix, R., Matzen, K., & Beech, C. (2013, Jun). Genetic control of Aedes mosquitoes. *Pathogens and Global Health, 107*(4), 170–179. doi:10.1179/2047773213Y.0000000095

Amouzou, A., Buj, V., Carvajal, L., Cibulskis, R., & Fergus, C. (2015). *Achieving the malaria MDG target: Reversing the incidence of malaria 2000–2015*. Geneva: World Health Organization.

Callaway, E. (2016). UK scientists gain licence to edit genes in human embryos. *Nature, 530*(7588), 18.

Carlson, D. F., Lancto, C. A., Zang, B., Kim, E. S., Walton, M., Oldeschulte, D., … Fahrenkrug, S. C. (2016, May 6). Production of hornless dairy cattle from genome-edited cell lines. *Nature Biotechnology, 34*(5), 479–481.

Charlesworth, C. T., Deshpande, P. S., Dever, D. P., Dejene, B., Gomez-Ospina, N., Mantri, S., … Porteus, M. H. Identification of pre-existing adaptive immunity to Cas9 proteins in humans. Retrieved from https://doi.org/10.1101/243345

Chneiweiss, H., Hirsch, F., Montoliu, L., Müller, A. M., Fenet, S., Abecassis, M., … Saint-Raymond, A. (2017, Oct). Fostering responsible research with genome editing technologies: a European perspective. *Transgenic Research*, *26*(5), 709–713.

Crispo, M., Mulet, A. P., Tesson, L., Barrera, N., Cuadro, F., dos Santos-Neto, P., … Menchaca, A. (2015). Efficient generation of myostatin knock-out sheep using CRISPR/Cas9 technology and microinjection into zygotes. *PLoS ONE*, *10*(8), e0136690.

de Lecuona, I., Casado, M., Marfany, G., Lopez Baroni, M., & Escarrabill, M. (2017, Dec 19). Gene editing in humans: Towards a global and inclusive debate for responsible research. *The Yale Journal of Biology and Medicine*, *90*(4), 673–681.

Doudna, J. A., & Charpentier, E. (2014, Nov 28). Genome editing. The new frontier of genome engineering with CRISPR-Cas9. *Science*, *346*(6213), 1258096. doi:10.1126/science.1258096.

Enserink, M. (2018, March 27). Interested in responsible gene editing? Join the (new) club. *Science*.

Egli, D., Zuccaro, M., Kosicki, M., Church, G., Bradley, A., & Jasin, M.. (2017, August 28). Interhomologue repair in fertilized human eggs? *BioRxiv*. Retrieved from https://doi.org/10.1101/181255

Fogarty, N. M. E., McCarthy, A., Snijders, K. E., Powell, B. E., Kubikova, N., Blakeley, P., … Niakan, K. K. (2017, Oct 5). Genome editing reveals a role for OCT4 in human embryogenesis. *Nature*, *550*(7674), 67–73.

Gantz, V. M., & Bier, E. (2015, Apr 24). Genome editing. The mutagenic chain reaction: a method for converting heterozygous to homozygous mutations. *Science*, *348*(6233), 442–444.

Grunwald, H. A., Gantz, V. M., Poplawski, G., Xu, X-ru S., Bier, E., & Cooper, K. L. (2018). Super-Mendelian inheritance mediated by CRISPR/Cas9 in the female mouse germline. *BioRxiv*. doi:10.1101/362558.

Haapaniemi, E., Botla, S., Persson, J., Schmierer, B., & Taipale., J.. (2018). CRISPR–Cas9 genome editing induces a p53-mediated DNA damage response. *Nature Medicine*, doi:10.1038/s41591-018-0049-z

Hainzl, S., Peking, P., Kocher, T., Murauer, E. M., Larcher, F., Del Rio, M., … Koller, U. (2017, Jul 13). COL7A1 Editing via CRISPR/Cas9 in recessive dystrophic epidermolysis bullosa. *Molecular Therapy*, pii: S1525-0016(17)30319-2.

Hammond, A., Galizi, R., Kyrou, K., Simoni, A., Siniscalchi, C., Katsanos, D., … Nolan, T. (2016, Jan). A CRISPR-Cas9 gene drive system targeting female reproduction in the malaria mosquito vector Anopheles gambiae. *Nature Biotechnology*, *34*(1), 78–83.

Hammond, A. M., Kyrou, K., Bruttini, M., North, A., Galizi, R., Karlsson, X., … Nolan, T. (2017, Oct 4). The creation and selection of mutations resistant to a gene drive over multiple generations in the malaria mosquito. *PLoS Genetics*, *13*(10), e1007039.

Hirsch, F., Lévy, Y., & Chneiweiss, H. (2017, Jan 4). CRISPR-Cas9: A European position on genome editing. *Nature*, *541*(7635), 30.

Huang, J., Chen, M., Whitley, M. J., Kuo, H. C., Xu, E. S., Walens, A., … Kirsch, D. G. (2017, Jul 10). Generation and comparison of CRISPR-Cas9 and Cre-mediated genetically engineered mouse models of sarcoma. *Nature Communications*, *8*, 15999. doi:10.1038/ncomms15999.

Hurlbut, J. B., Jasanoff, S., Saha, K., Ahmed, A., Appiah, A., Bartholet, E., … Woopen, C. (2018, Jul). Building capacity for a global genome editing observatory: Conceptual challenges. *Trends in Biotechnology*, *36*(7), 639–641.

Ihry, R. J., Worringer, K. A., Salick, M. R., Frias, E., Ho, D., Theriault, K., … Kaykas, A.. (2018). p53 inhibits CRISPR–Cas9 engineering in human pluripotent stem cells. *Nature Medicine*, doi:10.1038/s41591-018-0050-6.

Jain, A., Zode, G., Kasetti, R. B., Ran, F. A., Yan, W., Sharma, T. P., … Sheffield, V. C. (2017, Oct 17). CRISPR-Cas9-based treatment of myocilin-associated glaucoma. *Proceedings of the National Academy of Sciences of the United States of America*, *114*(42), 11199–11204.

Jansson, S. (2018, May 10). Gene-edited plants on the plate: the "CRISPR cabbage story". *Physiologia Plantarum*, doi:10.1111/ppl.12754.

Liang, P., Xu, Y., Zhang, X., Ding, C., Huang, R., Zhang, Z., … Huang, J. (2015). CRISPR/Cas9-mediated gene editing in human tripronuclearzygotes. *Protein & Cell*, *6*(5), 363–372.

Ma, H., Marti-Gutierrez, N., Park, S. W., Wu, J., Lee, Y., Suzuki, K., … Mitalipov, S.. (2017, Aug 24). Correction of a pathogenic gene mutation in human embryos. *Nature*, *548*(7668), 413–419. doi:10.1038/nature23305.

McGraw, E. A., & O'Neill, S. L. (2013). Beyond insecticides: new thinking on an ancient problem. *Nature reviews. Microbiology*, *11*(3), 181–193.

Mojica, F. J., Díez-Villaseñor, C., García-Martínez, J., & Soria, E. (2005, Feb). Intervening sequences of regularly spaced prokaryotic repeats derive from foreign genetic elements. *Journal of Molecular Evolution*, *60*(2), 174–182.

Mojica, F. J., Juez, G., & Rodríguez-Valera, F. (1993, Aug). Transcription at different salinities of Haloferax mediterranei sequences adjacent to partially modified PstI sites. *Molecular Microbiology*, *9*(3), 613–621.

Mojica, F. J. M., & Montoliu, L. (2016, Oct). On the origin of CRISPR-Cas technology: From prokaryotes to mammals. *Trends in Microbiology*, *24*(10), 811–820.

Montoliu, L. (2018). Transparency in animal experimentation. FEBS Network, Retrieved from https://network.febs.org/channels/728-viewpoints/posts/32348-transparency-in-animal-experimentation. Accessed on May 7, 2018.

Montoliu, L., Merchant, J., Hirsch, F., Abecassis, M., Jouannet, P., Baertschi, B., … Chneiweiss, H.. (2018). ARRIGE arrives: Toward the responsible use of genome editing. *The CRISPR Journal*, *1*(2). Published Online: 1 Apr https://doi.org/10.1089/crispr.2018.29012.mon

Montoliu, L., & Whitelaw, C. B. A. (2018). Unexpected mutations were expected and unrelated to CRISPR-Cas9 activity. *Transgenic Research*, *27*(4), 315–319.

Nelson, C. E., Hakim, C. H., Ousterout, D. G., Thakore, P. I., Moreb, E. A., Rivera, R. M. C., … Gersbach, A. (2016, Jan 22). In vivo genome editing improves muscle function in a mouse model of Duchenne muscular dystrophy. *Science*, *351*(6271), 403–407.

Saha, K., Hurlbut, J. B., Jasanoff, S., Ahmed, A., Appiah, A., Bartholet, E., … Woopen, C. (2018, Aug). Building capacity for a global genome editing observatory: Institutional design. *Trends in Biotechnology*, *36*(8), 741–743.

Schaefer, K. A., Wu, W.-H., Colgan, D. F., Tsang, S. H., Bassuk, A. G., & Mahajan, V. B. (2017). Unexpected mutations after CRISPR–Cas9 editing in vivo. *Nature methods*, *14*(6), 547–548.

Schaefer, K. A, Wu, W.-H., Colgan, D. F, Tsang, S. H, Bask, A. G., & Mahakam, V. B. (2018). Retraction: Unexpected mutations after CRISPR–Cas9 editing in vivo. *Nature Methods*, *15*, 394.

Seruggia, D., Fernández, A., Cantero, M., Pelczar, P., & Montoliu, L. (2015, May 26). Functional validation of mouse tyrosinase non-coding regulatory DNA elements by CRISPR-Cas9-mediated mutagenesis. *Nucleic Acids Research*, *43*(10), 4855–4867.

Smalley, E. (2018, June 6). As CRISPR–Cas adoption soars, summit calls for genome editing oversight. *Nature Biotechnology*.

Suzuki, K., Tsunekawa, Y., Hernandez-Benitez, R., Wu, J., Zhu, J., Kim, E. J., … Belmonte, J. C. I.. (2016). In vivo genome editing via CRISPR/Cas9 mediated homology-independent targeted integration. *Nature*, *540*(7631), 144–149. doi:10.1038/nature20565.

Whitworth, K. M., Rowland, R. R., Ewen, C. L., Trible, B. R., Kerrigan, M. A., Cino-Ozuna, A. G., … Prather, R. S. (2016). Gene-edited pigs are protected from porcine reproductive and respiratory syndrome virus. *Nature Biotechnology*, *34*(1), 20–22. doi:10.1038/nbt.3434

Wood, A. J., Lo, T. W., Zeitler, B., Pickle, C. S., Ralston, E. J., Lee, A. H., … Meyer, B. J. (2011, Jul 15). Targeted genome editing across species using ZFNs and TALENs. *Science*, *333*(6040), 307. doi:10.1126/science.1207773.

Yang, S., Chang, R., Yang, H., Zhao, T., Hong, Y., Kong, H. E., … Li, X. J. (2017, Jun 30). CRISPR/Cas9-mediated gene editing ameliorates neurotoxicity in mouse model of Huntington's disease. *The Journal of Clinical Investigation*, *127*(7), 2719–2724.

Yang, Y., Wang, L., Bell, P., McMenamin, D., He, Z., White, J., … Wilson, J. M. (2016, Mar). A dual AAV system enables the Cas9-mediated correction of a metabolic liver disease in newborn mice. *Nature Biotechnology*, *34*(3), 334–338.

Zetsche, B., Gootenberg, J. S., Abudayyeh, O. O., Slaymaker, I. M., Makarova, K. S., Essletzbichler, P., … Zhang, F. (2015, Oct 22). Cpf1 is a single RNA-guided endonuclease of a class 2 CRISPR-Cas system. *Cell*, *163*(3), 759–771.

DUAL USE IN NEUROSCIENTIFIC AND NEUROTECHNOLOGICAL RESEARCH: A NEED FOR ETHICAL ADDRESS AND GUIDANCE

James Giordano and Kathinka Evers

ABSTRACT

Extant and newly developing techniques and technologies generated by research in brain sciences are characteristically employed in clinical medicine. However, the increasing capabilities conferred by these approaches to access, assess and affect cognition, emotion and behavior render them viable and attractive for engagement beyond the clinical realm, in what are referred to as "dual-use" applications. Definitions of what constitutes dual-use research and applications can vary so as to include utilization in the public sector for lifestyle or wellness purposes – with growing participation of a do-it-yourself (i.e., biohacking) community, and an iterative interest and use in military and warfare operations. Such uses can pose risks to public safety, and challenge research ethics' principled imperative for non-harm (although while complete avoidance of any harm may be in reality impossible, certainly any/all harms incurred should be minimized). Thus, it is important to both clarify the construct of dual-use brain research and address the ethical issues that such research fosters. This chapter provides a review and clarification of the concept of dual-use brain science, and describes how current and emerging tools and techniques of brain research are actually or potentially employed in settings that threaten public health and incur ethical concerns. Key ethical issues are addressed, and recommendations for ethical guidance of potentially dual-use research are proposed.

Keywords: Dual-use; military; neuroscience; neurotechnology; neuroethics; policy; guidelines

Ethics and Integrity in Health and Life Sciences Research
Advances in Research Ethics and Integrity, Volume 4, 129–145

ISSN: 2398-6018/doi:10.1108/S2398-601820180000004009

INTRODUCTION: THE VIABILITY OF NEUROSCIENCE RESEARCH FOR DUAL-USE APPLICATIONS

Neuroscience employs a variety of methods and technologies to evaluate and influence neurologic substrates and processes of cognition, emotion, and behavior. In general, brain science can be either basic or applied research. Basic research focuses upon obtaining knowledge and furthering understanding of structures and functions of the nervous system on a variety of levels by employing methods of the physical and natural sciences. Applied research seeks to develop translational approaches that can be directly utilized to understand and modify the physiology, psychology, and/or pathology of target organisms, including humans. The techniques of both basic and applied neuroscience can be further categorized as those used to assess, and those used to affect the structures and functions of the nervous system, although these categories and actions are not mutually exclusive. For example, the use of certain ligands, toxins, and probes that are used to elucidate functions of various sites of the central and peripheral nervous system can also affect neural activity.

Neuroscience is broadly considered to be a natural and/or life science and there is implicit and explicit intent, if not expectation to develop and employ tools and outcomes of research in clinical medicine. Given the goals of medicine to elicit right and "good" treatment in patients' best interests, neuroscientific research is conducted in accordance with an undergirding maxim of non-harm (non-maleficence). However, absence of harm cannot always be assured for the use of research findings and/or products. This latter point has become somewhat contentious and is the focus of this chapter as regards the potential and actual uses of neuroscientific research that are distinct from intended applications, and/or specifically intended to incur demonstrably threatening consequences to individual and public health and/or environmental integrity. Such applications of scientific and technological research are referred to as "dual use."

DUAL USE: CONCEPTUAL ANALYSIS

Definitions

There are different definitions of dual use. Axiomatically, dual-use research refers to findings or products of scientific and/or technological studies that can be employed for more than one purpose. According to this definition, neuroscientific techniques, technologies, and information could be used for medical as well as non-medical (educational, occupational, lifestyle, military, etc.) purposes. Of particular note is that this formal, albeit general, definition of dual use does not indicate or suggest that such secondary uses incur burdens, risks, or harms beyond those anticipated for primary intent. Nor is it particularly useful, since everything that could be employed for more than one purpose would fall under "dual use." To reduce ambiguity and highlight potential threats of harm, the United States' National Institutes of Health Office of Science Policy (OSP) established the classification of "Dual Use Research of Concern" (DURC) that entails life science research that can be anticipated or expected to afford information, technologies, and/or products that can be engaged to incur deleterious consequences to public health and safety, agriculture, animals, environment, and/or national security. Intrinsic to this definition is the possibility that such research outcomes could be usurped to elicit harm (Office of Science Policy, 2012; 2014). As well, classification of DURC includes the use of tools and technologies that may pose risk and threat of harm as a consequence of inadvertent misuse (e.g., through laxity in laboratory containment, contamination, etc.). Of note is that although military and national security applications are certainly implied by, if not constituent to the OSP definition, and thus would warrant consideration and address, they are not specifically explicated.

A still more focused definition, which more stringently identifies such applications and aims, is provided by the European Commission, which classifies dual-use goods, products, and technologies as those "[…] normally used for civilian purposes, but which may have military applications" (European Commission Report, 2009). However, this definition does not either specify precisely which types of uses within the military would pose particular concerns that might be different from, or exacerbative to other occupational applications (e.g., cognitive, emotional, or behavioral alterations) that could pose risk or threat of harm. So, for example, would off-label use of neuropharmaceuticals or forms of noninvasive brain stimulation (NIBS) to optimize the performance of military personnel elicit different concerns given their potential engagement in national security, intelligence, or warfare operations? Here, while performance optimization represents a proximate goal, it could also be viewed as means instrumental to warfare (Eisenstein et al., 2017).

Of course, it could also be argued that such uses, performance enablements and resulting capabilities, could (and perhaps should) be used in intelligence and/or diplomatic operations to mitigate and subvert aggression, violence, and conflict (Giordano, 2015a). This remains a topic of ongoing debate. Of more focal concerns are uses of research findings and products to directly facilitate the performance of combatants, the integration of neural-machine interfaces to optimize combat capabilities of autonomous and/or semi-autonomous vehicles (e.g., drones) or modified insects or mammals ("cybernetic drones," e.g., the "DARPA beetle"), and development of biological and chemical weapons (i.e., neuroweapons). The potential for such uses is sustained by historical examples of military adoption of scientific and technological developments, dating at least to the middle of the nineteenth century (Moreno, 2012; Trumpener, 1975).

The increasing role of governmental support in both academic and industrial scientific enterprises during the early twentieth century fortified the establishment of unambiguous programs of military use of science and technology, inclusive of iterative developments in chemistry and biology that could be used to affect the nervous system (Christopher et al. 1999). Furthermore, given that a formal definition of a weapon is "a means of contending against others" (*Merriam Webster Dictionary*, 2008), it becomes difficult to specify whether and which neuroscientific tools and technologies, when employed in military contexts, can and should be regarded as weapons. Moreover, if a broad definition of dual use or DURC is exercised, then the criterion of individual or public safety or harm might necessitate a more granular address and analysis of offensive or defensive applications, questions of protection versus harm, and a more thorough exploration of means and ends, writ large. Lacking such conceptual clarification, categories of dual use and DURC could be considered vague and construed as either too broad or too narrow, which could incur practical as well as philosophical implications (Miller & Selgelid, 2008).

Thus, any ethical address or assessment may become problematic as differing criteria and specifications may sustain arguments about which research enterprises and outcomes should be monitored and regulated (Giordano, 2015a; Tennison et al., 2017). Therefore, an important step to uphold and advance the ethics of neuroscientific research would be a clarification and revision of the extant definitions of dual use and DURC. Toward this end, we propose the specification and distinction of those research enterprises that may yield results and products that 1) can be employed for extra-medical purposes within the public domain (in non-military occupational and lifestyle settings), and 2) those that are specifically employed for security, intelligence, military, and/or warfare operations.

Dual Use/DURC in Neuroscience and Neurotechnology

There is considerable literature addressing and describing evaluations and applications of neuroscientific tools' and techniques' capacity to sustain vigilance, increase coordination, improve memory and learning, decrease fatigue, and reduce stress. This has fostered steadily increasing interest in the use of such approaches to affect performance in certain occupational settings. Additionally, there is growing interest in employing neuroscientific techniques and technologies for educational as well as avocational/lifestyle (e.g., gaming, athletic) purposes. At present, most such applications are administered in supervised laboratory and/or clinical settings (inclusive of "off-label" medical uses), and are characteristically well monitored, in that distinct regulations apply for off-label use in research and medical practice.

For example, in research settings involving human subjects, the use of any/all drugs and devices, any research must comply with the mandates of the Declaration of Helsinki, and, notably, obtain:

- Approval by an institutional review board (IRB) or research ethics committee (REC) and, if the research engenders potential for serious risk to the health, safety, or welfare of a subject, approval of an investigational drug or device exemption (in the European Community by the European Medicines Agency).
- Informed consent from all subjects.
- Labeling of the drug and device for investigational use only.
- Monitoring of the study.
- Requisite records and reports.

In medical practice, the EC defines off-label uses as:

> Situations where a medicinal product is intentionally used for a medical purpose not in accordance with the authorized product information. Off-label use includes use in non-authorized paediatric age categories. Unless specifically requested, it does not include use outside the EU in an indication authorized in that territory which is not authorized in the EU. (European Medical Agency, 2015, 2016).

This definition establishes that drugs and devices that have European marketing authorization will be eventually considered "off-label", while those products without this authorization will be regarded as "unlicensed." These definitions and existing regulations presume that any and all off-label use represents a matter of medical judgment and occurs in a conscientious manner with regard to good clinical practices (Lenk & Duttge, 2014).

Do-it-Yourself/Neurobiohacking

However, there is also a growing do-it-yourself (DIY)/biohacking community that is dedicated to altering commercially available direct-to-consumer (DTC) products to perform different functions, and/or creating new products capable of affecting neurobiological functions. Biohacking can be articulated in three research domains: synthetic biology (e.g., genetic and molecular editing); biochemistry (e.g., development of neurotropic agents that can be used either singularly or in chemical cocktails); and biotechnology (human–machine interfaces, technological implants, and prosthetics). These categories and their products are not mutually exclusive.

Biohacking typically implies modifications for benevolent ends (i.e., "white-hat" hacking), inclusive of development of agents and devices to improve human cognition, emotion, and behavioral performance. DIY scientists/biohackers often work in coordination within an informally organized community, and much of their research is made publicly available through open access databases and websites of community laboratories. The spirit of the DIY/biohacking community reflects a movement to make biology "easier to engineer," and more publicly accessible and available (see, e.g., https://www.kickstarter.com/.../neuroscience-for-the-people-bring-diy-science-to-the-people). In part, this is constituent to an expanding trend toward "open source" biology that has influenced both research institutions and the public. Additionally, "open source" biology has captured an economic market niche: engineered and modified organisms, drugs, and devices can be sold; community laboratories can be purchased (by conventional commercial entities); and both community laboratories and individual DIY biohackers can be subsidized through venture capital. With manuals and methods available online, it is relatively easy to establish and run a laboratory, and interested individuals and groups can obtain guidance on producing and/or manipulating a variety of neurobiological techniques and technologies. But there is also a "black-hat" hacking community that engages DIY approaches to neurobiology to produce pathogens, or to incur other disruptions in individual or community stability and safety.

Both "white hat" and clearly "black hat" DIY neuroscience can pose regulatory, health, and security risks (Fitz & Reiner, 2015). Independent laboratories and researchers do not always abide by the comprehensive policies that academic and industrial research entities must follow. As well, there is increasing use of the "dark web" (covertly accessed Internet) by both "white hat" and "black hat" biohackers to facilitate exchange of information in ways that impede surveillance. This community presents particular dual-use research concerns in that 1) outcomes and products may be used or misused in ways that adversely impact upon individual and public health and safety, as well as the integrity of flora and fauna in the environment; 2) limitations and/or lassitude in research practices and/or laboratory conditions may incur accidental release of information or products that can pose health and environmental risks and harms; and/or 3) activities may be subsidized and outcomes and products utilized by national and non-state venture capitalists with explicit intent toward disrupting public safety, stability, and health (DiEuliis & Giordano, 2017).

These possibilities evoke security concerns on local, national, and international scales, and have warranted involvement of crime prevention and public safety agencies (e.g., the United States Federal Bureau of Investigation) to establish dialog with, and insight to the DIY biohacking community. What is important to note is that neuroscientific and neurotechnological research and development is occurring on a variety of levels (from large-scale academic and industrial laboratories to individual DIY experimenters) and is becoming increasingly international. This increases the possibility for dual-use research and DURC, and generates questions about what constitutes research for security purposes (i.e., preparatory defense) versus military/warfare (i.e., offensive capability) purposes (Giordano, 2016a; Palchick et al, 20917).

MILITARY/WARFARE DUAL-USE APPLICATIONS OF NEUROSCIENTIFIC AND NEUROTECHNOLOGICAL RESEARCH

The use of neuroscience and technology for military/warfare purposes is realistic and represents a clear and present concern. Illustratively, a 2008 report by the *ad hoc* Committee on Military and Intelligence Methodology for Emergent Neurophysiological and Cognitive/Neural Science Research in the Next Two Decades by the National Research Council of the United States Academy of Sciences claimed that neuroscience and technology, while possessing potent capabilities, were not as yet demonstrably employable in military operations (National Research Council of the National Academies of Science, 2008). However, by 2014, the Committee's subsequent report asserted that neuroscience and technology had matured considerably and were being increasingly considered, and in some cases evaluated for operational use in security, intelligence, and defense operations (National Research Council of the National Academies of Science, 2014). This evaluation was seconded by a 2013 Nuffield Council Report, and is supported by a series of white papers by the Strategic Multilayer Assessment (SMA) Group of the Joint Staff of the Pentagon that illustrated the viability and value of the brain sciences to security, intelligence, and military operations (for overviews of, and access to, reports, see http://nsi-team.com/sma-publications/).

In large part, the iterative recognition of the viability of neuroscience and technology in these agenda reflects the pace and breadth of developments in the field. Although a number of nations have pursued, and are currently pursuing neuroscientific research and development for military purposes, perhaps the most proactive efforts in this regard have been conducted by the United States Department of Defense, with most notable and rapidly maturing research and development conducted by the Defense Advanced Research Projects Agency (DARPA) and Intelligence Advanced Research Projects Activity (IARPA). To be sure, many DARPA projects are explicitly directed toward advancing neuropsychiatric treatments and interventions that will improve both military and civilian medicine (e.g., Systems-based Neurotechnologies for Emerging Therapies – SUBNETS; Restoring Active Memory – RAM; etc.; see: www.darpa.mil, for overview). While such uses can give rise to neuroethico-legal and social issues, the possibility of employing the techniques and technologies generated in these research programs for biological augmentation can generate unease and additional ethical questions if and when construed in contexts of the performance of combat personnel (Giordano, 2015a; Moreno, 2012; Tennison et al, 2017). As noted in the 2008 National Research Council report, "[…] for good or for ill, an ability to better understand the capabilities of the body and brain […] could be exploited for gathering intelligence, military operations, information management, public safety and forensics" (National Research Council of the National Academies of Science, 2008). However, definitions of "good" may vary and what is regarded as good for some may present harm to others.

The potential for neuroscience and technology to afford insight, understanding, and capability to affect cognitive, emotional, and behavioral aspects of individuals and groups render the brain sciences particularly attractive for use in security, intelligence, and military/warfare initiatives. To approach this issue, it is important to establish four fundamental premises. First, neuroscience and its technologies are and will be increasingly and more widely incorporated into approaches to national security, intelligence gathering and analysis, and aspects of military operations. Second, such capabilities afford considerable power. Third, many countries are actively developing and subsidizing neuroscientific and technological research under dual-use agendas or for direct incorporation into military programs (see Tennison et al, 2017, for overview). Fourth, as noted in both 2008 and 2014 National Academies' reports, these international efforts could lead to a "capabilities race" as nations react to new developments by attempting to counter and//or improve upon one another's discoveries. This type of escalation represents a realistic possibility with potential to affect international security (Dando, 2007, 2014, 2015; Giordano et al., 2010; McCreight, 2015). Such "brinksmanship" must be acknowledged as a potential impediment to attempts to develop ethical analyses and guidelines (that inform or prompt policies) that seek to constrain or restrict these avenues of research and development (Forsythe & Giordano, 2011; Tractenberg et al., 2015).

A brief discussion of current developments and domains of neuroscientific and neurotechnological research and applications that are being used or considered for military and warfare operations is provided below.

Neural Systems Modeling and Human–Machine Interactive Networks in Intelligence, Training, and Operational Systems

Research in cognitive and computational neuroscience is being engaged to improve:

- Human cognitive performance – through improved understanding of basic processes involved in memory, emotion, and reasoning to support and enhance intelligence analysis, planning and forecasting capabilities.
- Training efficiency – by using knowledge and tools of cognitive neuroscience to enable more rapid acquisition and mastery of knowledge and skills with more durable retention.
- Team process performance – via engagement of systems engineering of human/brain –interfacing to enhance information processing capability of individuals, organizations, and surveillance and weapons' systems (drones). Research in this domain generally employs a technology readiness/technology transfer approach that utilizes a nine-level assessment and articulation scheme (from observation of basic principles, through evaluation and validation in a relevant environment, to full operational readiness) to advance research, development, testing, and evaluation, toward rapid use. At present, a number of human/brain-machine interfaces are transitioning from development through test and evaluation stages toward operational readiness within a five-year cycle (for review, see: Oie & McDowell, 2015; Stanney et al. 2015).

Neuroscientific Technological Approaches to Optimizing Performance and Resilience in Combat and Military Support Personnel

Neurocognitive studies employing various forms of neuroimaging, neurogenomics, proteomics, and biomarker assessment are being used to identify and define neural networks involved in several dimensions of operational performance of military combat and support personnel. These approaches seek to identify and isolate neural structures, systems, and functions that can be "targeted" for interventions utilizing noninvasive brain stimulation, pharmacological agents (e.g., stimulants, eugeroics; nootropics), or cognitive- behavioral training to facilitate, sustain, and/or improve performance capability and reduce dysfunction.

These assessments and interventions have been and could be regarded as components of preventive military medicine (Abney, Lin, & Mehlman, 2015; Giordano, 2015a, 2017a). Moreover, studies conducted within and/or directly funded by the military have been utilized for their "reverse dual-use" applications in civilian occupational and preventive medical contexts. However, these techniques and technologies also raise concerns about creating "super soldiers" that obtain fortified cognitive, emotional, and behavioral characteristics that maximize their combat capabilities. A contrary position posits that such methods could, and arguably should be engaged to instead produce soldiers who possess improved decision-making, interpersonal, and perhaps even empathic characteristics and skills (Wurzman & Giordano, 2015). These contrasting views fuel current discussion and debate.

Direct Weaponization of Neuroscience and Neurotechnology

The formal definition of a weapon as "a means of contending against others" can be extended to include any implement or technique that can be "[...] used to injure, defeat, or destroy" (*Merriam Webster Dictionary*, 2008). Both definitions apply to products of neuroscientific and neurotechnological research that can be employed in military/warfare scenarios. The objectives for neuroweapons in a traditional military/warfare context (e.g., combat) may be achieved by augmenting or degrading functions of the nervous system, so as to affect cognitive, emotional, and/or motor activity and capability (e.g., perception, judgment, morale, pain tolerance, or physical abilities and stamina) necessary for combat. Many technologies (e.g., neurotropic drugs; neurostimulatory devices) can be employed to produce these effects, and there is demonstrated utility for neuroweapons in both conventional and irregular warfare scenarios (Dando, 2014, 2015; Giordano, 2017a; Moreno, 2012; Wurzman & Giordano, 2015).

To be sure, the weaponized use of neuroscientific tools and products is not new. Historically, such weapons have included nerve gas, and various drugs (British Medical Association, 2007; Gross, 2010; Szinicz, 2005). Weaponized gas has taken several forms: lachrymatory agents (tear gases), toxic irritants (e.g., phosgene, chlorine), vesicants (blistering agents; e.g., mustard gas), and paralytics (e.g., sarin). Pharmacologic stimulants (e.g., amphetamines) and various ergogenics (e.g., anabolic steroids) have been used to augment performance of combatants (Kamienski, 2016) and sedatives (e.g., barbiturates) have been employed to enhance cooperation during interrogation (Moreno, 2012). Sensory stimuli (e.g., high intensity sound; prolonged flashing lights; irritating music or noise) have been applied as neuroweapons to incapacitate the enemy, and even sleep deprivation and distribution of emotionally provocative information in psychological operations (i.e., PSYOPS) could rightly be regarded as forms of weaponized applications of neuroscientific and neurocognitive research.

At present, outcomes and products of computational neuroscience and neuropharmacologic research could be used for more indirect applications, such as enabling human efforts by simulating, interacting with and optimizing brain functions, and the classification and detection of human cognitive, emotional, and motivational states to augment intelligence, or counter-intelligence tactics. Human/brain-machine interfacing neurotechnologies capable of optimizing data assimilation and interpretation systems by mediating access to and manipulation of signal detection, processing, and/or integration are being explored for their potential to de-limit "human weak links" in the intelligence chain (Giordano and Wurzman, 2016).

Additionally, there is interest in employing neurotechnology to augment the role, capability, and effects of psychological operations (PSYOPS) in military and political missions. As noted in several SMA Group reports to the Pentagon, the intent and desired outcomes of this research is an improved understanding of neural bases and effects of narratives that can afford insights to influences and processes that affect brain development, function, and behavior, which can be operationalized to mitigate violence on a variety of scales (Cabayan, 2012; Cabayan & Canna, 2014; Giordano, 2014; Giordano, 2016b; Giordano, Casebeer, & DiEuliis, 2013; Giordano & DiEuliis, 2015).

While such projects may give rise to concerns about the use of neuroscientific and neurotechnological research to develop methods of "mind reading" (Evers & Sigman, 2013) and "mind control," greater (if not greatest) concerns are generated by neuropharmacologic, neurotoxicologic, neuromicrobiologic, and neurotechnologic research that has potential to develop non-lethal or lethal weapons in combat-related and/or special operations' deterrence operations (British Medical Association, 2007; Dando, 2014, 2015; Giordano & Wurzman, 2011; Nuffield Council, 2013; Wheelis, 2003; Wurzman & Giordano, 2015). Weaponizable products of neuroscientific and neurotechnological research can be utilized to affect 1) memory, learning, and cognitive speed; 2) wake-sleep cycles, fatigue, and alertness; 3) impulse control; 4) mood, anxiety, and self-perception; 5) decision-making; 6) trust and empathy, and 7) movement and performance (e.g., speed, strength, stamina, motor learning, etc.). In military/warfare settings, modifying these functions can be employed to mitigate aggression and foster cognitions and emotions of affiliation or passivity; induce morbidity, disability, or suffering and "neutralize" potential opponents; or incur mortality.

Non-lethal and lethal neuroweapons include various categories and classes of psycho-neuroactive drugs, a variety of microbial agents (e.g., bacterial and viral strains) that act directly or exert effect upon the central and/or peripheral nervous system; organic toxins; and neurotechnological devices (e.g., sensory and brain stimulation approaches) and products (e.g., nanotechnologically derived substances). Additionally, there is concern that brain-machine interfacing and neural network-derived computational decision systems could be employed to develop remote control or autonomous/semi-autonomous capability for unmanned aerial and ground vehicles and/or organisms (e.g., insects, small mammals) that could function as weapon platforms. The use of unmanned vehicles as weapons is not novel, and the realization of fully autonomous capability is still lacking. But the iterative progression and integration of neurotechnologically enabled capabilities render such weapons increasingly viable, and therefore a source of trepidation about near-term future developments that could be generated from ongoing research in neural architectures' and human-machine systems. For reviews of the historicity, current development, and ethical issues of autonomous and semi-autonomous weapons, see Asaro (2006, 2008); Danielson (2011); and Howlader and Giordano (2013).

ETHICAL ADDRESS AND OVERSIGHT

Extant Treaties and Regulations

In the EU, United States, and most Western countries, DURC is proscribed by a number of legally binding international treaties and non-binding agreements.

An analytic overview of the issues they address is provided by Kuhlau et al. (2008, 2012). Notable among those legally binding entities are the Hague Convention(s); Geneva Protocol; Biologic, Toxin and Weapons Convention (BTWC); Chemical Weapons Convention (CWC); United Nations' Security Council Resolution 1540; Lisbon Treaty (EC Regulation 428/2009); and the 2014 Arms Trade Treaty, which seek to restrict or constrain research programs (and the transnational exchange) of defined microbiological agents, organic toxins, and drugs. However, the development, relative ease of acquisition, and possible use of gene editing tools, such as CRISPR/Cas-9, may enable modification of biological entities not currently covered by these conventions and treaties (e.g., microbes, agricultural products, insects), to be employed as weapons (DiEuliis and Giordano, 2017). So, while some domains of brain science fall within the scope of the current BTWC, CWC and other treaties and policies, others (such as neurotechnologies and the aforementioned genetically modified agents), do not (Dando, 2007, 2014, 2015; Gerstein & Giordano, 2017; Giordano, 2015, 2016b, 2017b). As well, the use of certain drugs (e.g., sedative, tranquilizing, and analgesic agents) as "calmatives" and sensory disruption devices (e.g., sonic generators) in civil law enforcement is not addressed or restricted by current treaties and agreements.

Thus, as brain science continues to advance and to be assessed for dual uses, it will be important to address: 1) whether and to what extent techniques and technologies used in and produced by neuroscientific research can incur risks and harms; 2) whether newly developing neuroscientific methods and products fall under the purview of current treaties; and if they do not 3) what revisions would be necessary to create a more realistic and inclusive orientation to DURC. These latter two points were explicated by the Australia Group at the Eighth Review Conference of the parties of the BTWC in December 2016. But international treaties and policies do not guarantee cooperation. Moreover, products that are studied and intended for the health/medical market may fall under such treaties' exemptions for biomedical experimental or clinical purposes, and these products can spill-over to military use, and/or, as previously noted, may be utilized by do-it-yourself (DIY)/biohackers working as non-state actors (Dando, 2015; Giordano & Wurzman, 2011; Giordano, 2016b, 2017b; National Research Council, 2008, 2014; Wurzman & Giordano, 2015).

Ethical Address and Guidance

In this age of increasingly open science, it is therefore important to recognize that almost any neuroscientific and neurotechnological information, method(s) or technologies could be usurped into dual-use and explicit military and/or warfare agendas. Awareness of this potential is varied: a recent study reported that neuroscientists were more likely to consider colleagues' work as being more apt to pose dual-use risk or likelihood than their own (Kosal and Huang, 2015). Furthermore, dual-use applications are often distinct from the intent of the research as proposed and/or conducted, and, in some cases, they are beyond the focus of the scientists conducting the work. This makes attempts to address and prescribe methods toward ethical address, guidance, and governance difficult (Kuhlau et al., 2012).

To be sure, we do not here attempt to address or propose methods to regulate the actual uses of neuroscientific information, techniques, and technologies, as this would be beyond our scope, and almost impossible given the myriad variables involved. The information and products of brain science can and likely will be employed for each and all of the uses described above, inclusive of various national security, military and warfare operations. Therefore, our emphasis is to raise awareness and foster recognition of potential and likely dual-use(s) that particular (if not any/all) research projects may incur. As consistent with a guiding ethical principle of non-harm, such a stance would address and acknowledge both intentional and unintended consequences and risks evoked by dual use of neuroscientific and neurotechnological research. However, here the question is whether, and to what extent researchers are morally obligated to engage pre-emptive caution to avoid or prevent undesirable future outcomes of their research (Chameau, Balhaus, et al. 2014; Dando, 2007; Kuhlau et al. 2012; Miller and Selgelid, 2008).

Recently, a concerned group of neuroscientists developed a document to establish a pledge to proscribe active participation in brain research that is explicitly to be used for military purposes (Bell, 2015). This is laudable, but competitive pressures of extramural funding may make such a stance difficult to sustain, unless there is some form of generalized administrative conformity that guides research and researchers' engagement of projects with dual-use capability. Toward this end, we propose the following:

- Establishment of durable working groups dedicated to ongoing address, and assessment of research activities with potential for DURC/military-warfare uses.
- Engagement of (members of) this working group in proactive discourse with relevant interest groups (e.g., the public; national defense and security organizations/councils; commercial entities; representatives from the DIY/neurobiohacking community).
- Development of professional and public educational programs that provide ongoing seminars/symposia, webinars, publications, and online informational material addressing DURC/military-warfare use of research.
- Formulation of updated mission statements for consensus/ratification that affirms the intent of any/all research for civilian biomedical purposes, and proscribes intent for specific dual use(s) of concern, inclusive of military-warfare applications.
- Establishing requirements that any/all research personnel:
  - participate in these educational activities on a defined, regular basis;
  - explicitly affirm intent of purpose (through signatory affirmation of a representative statement of intent); and
  - explicate such intent in a formal statement to be included on any/all publications and proposals.

However, this stance does not necessarily address the possibility for other avenues of dual use (e.g., DIY science and technology), and potential risks and harms that these may incur. Therefore, while an important and perhaps necessary first step could be to restrict/ban military support of any/all projects within and/or directly derived from neuroscientific research efforts, this is not sufficient. Vital next steps include evaluation of ongoing and planned research projects; assessment of overt and tacit aspects that foster dual use and possible risks and harms incurred by the information and/or products these studies will generate; and estimation/determination of the probability and timeline of such dual-use capabilities (Giordano, 2016a).

Thus, it is practical to determine which efforts are most likely for dual use, and therefore of highest priority for ethical address. Timelines of dual-use capability can be estimated by assessing the current maturity and anticipated maturational trajectory of particular neuroscientific and/or neurotechnological developments, contingencies for maturity to fruition and operationalization in defined contexts, and plotting these against (a) defined level of control in provision and access of use; and (b) relevant markets' influence; that is, demand and relative power to affect access and use (Schnabel et al., 2014) This timeline can then be fitted to estimations and probabilities of relative risks and harms, both in general and in defined circumstances, such as the possibility for inapt use/misuse in the public domain; "runaway effects" of science and technology; use in military/warfare settings, etc. (Giordano et al., 2014). This is important to enable focus upon those research enterprises that pose greatest possibility and probability for near- to mid-future, and mid- to high-risk dual use.

In light of these current and near-future risks, a neuroethico-legal and social risk-assessment and mitigation paradigm has been developed that is based upon recommendations of the Royal Society Working Group on Neuroscience (2012), the United States' National Research Council (2014), and those proposed by Casebeer (2014). This approach calls for responsibility for accurate assessment of research, proactive responsiveness and engagement by researchers, revision of identified research efforts that pose concern, and efforts toward regulation. Working within this approach, specific queries and framing constructs have been proposed to enable more accurate determination of if and how specific uses of neuroscience and neurotechnological research give rise to ethical issues (Giordano, 2012, 2015b, 2017b).

CONCLUSION

Despite such recommendations for dual-use research, two important realities must be considered: First, almost any research could be taken up and used for purposes not explicit to intent; and second, it is likely that restricting university-based research with viability for dual or direct (military/warfare) use will prompt expansion/foster (new) research enterprises within the military (as in Germany, for example). Thus, such uses of brain science will not "go away." Therefore, it will be important and necessary to support education about such uses, and engage forums that can oversee and govern the use of neuroscience and technology in these ways.

We must assume that military research would axiomatically be directed toward studying and developing techniques and tools that could be employed in operational settings (e.g., intelligence, human terrain campaigns, warfare). By implication, military ethics would likely be viewed as most relevant and applicable to guiding such use. But this prompts questions and debate about 1) whether extant "grounding principles," either in or derived from accepted constructs about "just" conduct of conflict (e.g., *jus ad bellum*; *jus in bello*; *jus contra bellum*) are sufficiently detailed and developed to remain apace with emerging capabilities in brain science (British Medical Association, 2016; Eisenstein, Naumann et al., 2017; Tennison et al., 2017; Tractenberg, FitzGerald, Giordano, 2015) and 2) if and how such science – and ethics – informs and directs revision of current (and/or formulation of new) treaties and conventions that govern biological agents in warfare (Gerstein & Giordano, 2017).

Regardless of whether focal to research or use-in-practice, and irrespective of the actual system(s) of ethics employed, any such address and oversight must be forward-looking, descriptive, conceptually informed, predictive, preparatory, analytic, and precautionary for contingencies and exigencies that can occur as neuroscientific and neurotechnological research – and the environments in which it may be used – evolve. We acknowledge the complexity and magnitude of the endeavor and efforts required, yet advocate dedication to such enterprise as worthwhile. For the question is not whether dual use of brain research will occur, but rather when, in what ways, to what extent, and to what effect? The answers to these questions are a responsibility borne – to a great extent – by the scientific community.

REFERENCES

Abney, K., Lin, P., & Mehlman, M. (2015). Military neuroenhancement and risk assessment. In J. Giordano (Ed.), *Neurotechnology in National Security and Defense: Practical Considerations, Neuroethical Concerns* (pp. 239–248), Boca Raton, FL: CRC Press.

Asaro, P. M. (2006). What should we want from a robot ethic? *Int Rev Information Ethics*, *6*(6), 9–16.

Asaro, P. M. (2008). How just could a robot war be? In A. Briggle, K. Waelbers, & P. A. E. Brey (Eds.), *Current Issues in Computing and Philosophy* (pp. 50–64). Amsterdam: IOS Press.

Bell, C. (2015). Why neuroscientists should take the pledge: A collective approach to the misuse of neuroscience. In J. Giordano (Ed.), *Neurotechnology in National Security and Defense: Practical Considerations, Neuroethical Concerns* (pp. 227–238). Boca Raton, FL: CRC Press.

British Medical Association. (2007). *The Use of Drugs as Weapons*. London: BMA Press.

British Medical Association. (2016). *Armed Forces Ethics Tool Kit*. London: BMA Press.

Cabayan, H. (Eds.). (July 2012). *National security challenges: Insights from social, neurobiological and complexity sciences*. Department of Defense; Strategic Multilayer Assessment Group – Joint Staff/J-3/ Pentagon Strategic Studies Group.

Cabayan, H., & Canna, S. (Eds.). (December, 2014). *Multi-method assessment of ISIL*. Department of Defense; Strategic Multilayer Assessment Group-Joint Staff/J-3/Pentagon Strategic Studies Group.

Casebeer, W. D. Plenary testimonial presented to US Presidential Commission for the study of bioethical issues. (2014). Retrieved from www.bioethics.gov/node/2224. Accessed on March 18, 2017).

Chameau, J. L., Balhaus, W. F., & Lin, H. S. (2014). *Emerging and readily available technologies and national security: A framework for addressing ethical, legal, and societal issues*. Washington, DC: National Academies Press.

Christopher, G. W., Cieslak, T. J., Pavlin, J. A., & Eitzen, E. M. (1999). Biological warfare: A historical perspective. In J. Lederburg (Ed.), *Biological warfare: Limiting the threat* (pp. 17–36). Cambridge, MA: MIT Press.

Dando, M. (2007). *Preventing the future military misuse of neuroscience*. New York, NY: Palgrave-Macmillan.

Dando, M. (2014). Neuroscience advances in future warfare. In J. Clausen & N. Levy (Eds.), *Handbook of Neuroethics*. Dordrecht: Springer.

Dando, M. (2015). *Neuroscience and the future of chemical-biological weapons*. New York, NY: Palgrave-Macmillan.

Danielson, P. (2011). Engaging the public in the ethics of robots for war and peace. *Philosophy and Technology*, *24*(3), 239–249.

DiEuliis, D., & Giordano, J. (2017). Whygene editors like CRISPR/Cas may soon be a game-changer for neuroweapons. *Health Security*, *15*(3), 296–302.

Eisenstein, N., Naumann, D., Burns, D., Stapley, S., & Draper, H. (2017). Left of Bang interventions in trauma: Ethical implications for military medical prophylaxis. *Journal of Medical Ethics*, *44*(7), 125–131.

Evers, K., & Sigman, M. (2013). Possibilities and limits of mind-reading: A neurophilosophical perspective. *Consciousness and Cognition*, *22*(3), 887–897.

European Commission, European Regulation (EC). (2009). No 428/2009. Retrieved from http://ec.europa.eu/trade/import-and-export-rules/export-from-eu/dual-use-controls/

European Medical Agency. (2015, 2016). Regulatory standing of direct marketing of medicines and medical products. Retrieved from http://www.ema.europa.eu/ema/docs/en_GB/document_library_000114.jsp. Accessed on March 25, 2017.

Fitz, N., & Reiner, P. (2015). The challenge of crafting policy for do-it-yourself brain stimulation. *Journal of Medical Ethics*, *41*(5), 410–412.

Forsythe, C., & Giordano, J. (2011). On the need for neurotechnology in the national intelligence and defense agenda: Scope and trajectory. *Synesis: A Journal of Science, Technology, Ethics and Policy*, *2*(1), 5–8.

Gerstein, D., & Giordano, J. (2017). Re-thinking the biological, toxin and weapons convention? *Health Security*, *15*(6), 638–641.

Giordano, J. (2012). Neurotechnology as demiurgical force: Avoiding Icarus' folly. In J. Giordano (Ed.), *Neurotechnology: Premises, Potential and Problems* (pp. 1–14). Boca Raton, FL: CRC Press.

Giordano, J. (Ed.). (May 2014). *Leveraging neuroscience and neurotechnological (neuroS/T) ddevelopment with focus on influence and deterrence in a networked world*. Department of Defense; Strategic Multilayer Assessment Group- Joint Staff/J-3/ Pentagon Strategic Studies Group.

Giordano, J. (2015a). Neurotechnology, global relations and national security: Shifting contexts and Neuroethical demands. In J. Giordano (Ed.), *Neurotechnology in National Security and Defense: Practical Considerations, Neuroethical Concerns* (pp. 1–10). Boca Raton, FL: CRC Press.

Giordano, J. (2015b). A preparatory neuroethical approach to assessing developments in neurotechnology. *AMA J Ethics*, *17*(1), 56–61.

Giordano, J. (2016a). The neuroweapons threat. *Bull Atomic Scientists*, *72*(3), 1–4.

Giordano, J. (Ed.) (2016b, March), *A biopsychosocial science approach for understanding the emergence of and mitigating violence and terrorism*. Department of Defense; Strategic Multilayer Assessment Group-Joint Staff/J-3/Pentagon Strategic Studies Group.

Giordano, J. (2017a). Battlescape brain: Engaging neuroscience in defense operations. *HDIAC Journal*, *3*(4), 13–17.

Giordano, J. (2017b). Toward an operational neuroethical risk analysis and mitigation paradigm for emerging neuroscience and technology (neuroS/T). *Experimental Neurology*, *287*(4), 492–495.

Giordano, J., Casebeer, W., & DiEuliis, D. (Eds.) (2013, April), *Topics in operational considerations on insights from neurobiology on influence and extremism*. Department of Defense; Strategic Multilayer Assessment Group-Joint Staff/J-3/ Pentagon Strategic Studies Group.

Giordano, J., Casebeer, W. D., & Sanchez, J. (2014). *Assessing and managing risks in systems neuroscience research and its translation: A preparatory neuroethical approach*. Washington, DC: Paper presented at 6th Annual Meeting of the International Neuroethics Society.

Giordano, J., & DiEuliis, D. (Eds.). (May, 2015). Social and Cognitive Neuroscientific Underpinning of ISIL Behavior, and Implications for Strategic Communication, Messaging and Influence. Department of Defense; Strategic Multilayer Assessment Group-Joint Staff/J-3/Pentagon Strategic Studies Group.

Giordano, J., Forsythe, C., & Olds, J. (2010). Neuroscience, neurotechnology and national security: The need for preparedness and an ethics of responsible action. *AJO B-Neuroscience*, *1*(2), 1–3.

Giordano, J., & Wurzman, R. (2011). Neurotechnologies as weapons in national intelligence and defense – an overview. *Synesis: A Journal of Science, Technology, Ethics and Policy*, *2*(1), 138–154.

Giordano, J., & Wurzman, R. (2016). Integrative computational and neurocognitive science and technology for intelligence operations: Horizons of potential viability, value and opportunity. *STEPS- Science, Technology, Engineering and Policy Studies*, *2*(1), 34–38.

Gross, M. L. (2010). Medicalized weapons and modern war. *Hastings Center Report*, *40*(1), 34–43.

Howlader, D., & Giordano, J. (2013). Advanced robotics: Changing the nature of war and thresholds and tolerance for conflict – implications for research and policy. *Journal of Philosophy, Science & Law*, *13*(13), 1–19.

Kamienski, L. (2016). *Shooting up: A short history of drugs and war*. Oxford: Oxford University Press.

Kosal, M. E., & Huang, J. Y. (2015). Security implications and governance of cognitive neuroscience. *Politics Life Sci*, *34*(1), 93–108.

Kuhlau, F., Eriksson, S., Evers, K., & Höglund, A. (2008). Taking due care: Moral obligations in dual use research. *Bioethics*, *22*(9), 477–487.

Kuhlau, F., Höglund, A., Evers, K., Höglund, A., & Eriksson, S. (2009). A precautionary principle for dual use research in the life sciences. *Bioethics*, *25*(1), 1–8.

Kuhlau, F., Evers, K., Eriksson, S., & Höglund, A. (2012). Ethical competence in dual use life science. *Applied Biosafety*, *17*(3), 120–127.

Lenk, C., & Duttge, G. (2014). Ethical and legal framework and regulation for off-label use: European perspective. *Therapeutics and Clinical Risk Management*, *10*, 537–546.

McCreight, R. (2015). Brain brinksmanship: Divising neuroweapons looking at battlespace, doctrine and strategy. In J. Giordano (Ed.), *Neurotechnology in National Security and Defense: Practical Considerations, Neuroethical Concerns* (pp. 115–132). Boca Raton, FL: CRC Press.

Merriam-Webster's Dictionary. (2008). 'Weapon'. Retrieved from http://www.miriam-webster.com/dictionary/weapon. Accessed on March 25, 2017.

Miller, S., & Selgelid, M. (2008). *Ethical and philosophical considerations of the dual-use dilemma in biological sciences*. New York, NY: Springer.

Moreno, J. D. (2012). *Mind wars*. New York, NY: Bellevue Press.

National Research Council of the National Academies of Science. (2008). *Emerging cognitive neuroscience and related technologies*. Washington, DC: National Academies Press.

National Research Council of the National Academies of Science. (2014). *Emerging and readily available technologies and national security: A framework for addressing ethical, legal and societal issues*. Washington, DC: National Academies Press.

Nuffield Council on Bioethics. (2013). *Novel neurotechnologies: Intervening in the brain*. London: Nuffield Council on Bioethics.

Office of Science Policy (OSP). (2012). *United States government policy for institutional oversight of life sciences dual use research of concern (September 2014)*. Washington, DC: US Government Printing Office.

Oie, K. S., & McDowell, K. (2015). Neurocognitive engineering for systems' development. In J. Giordano (Ed.), *Neurotechnology in National Security and Defense: Practical Considerations, Neuroethical Concerns* (pp. 33–50). Boca Raton, FL: CRC Press.

Schnabel, M., Kohls, N. B., Sheppard, B., & Giordano, J. (2014). *New paths through identified fields: Imaging domains of neuroethical, legal and social issues of global use of neurotechnology by qualitative modeling and probabilistic plotting within a health promotions paradigm.* Washington, DC: Paper presented at 6th Annual Meeting of the International Neuroethics Society.

Stanney, K. M., Hale, K. S., Fuchs, S., Baskin-Carpenter, A., & Berka, C. (2015). Neural systems in intelligence and training applciaitons. In J. Giordano (Ed.), *Neurotechnology in National Security and Defense: Practical Considerations, Neuroethical Concerns* (pp. 23–32). Boca Raton, FL: CRC Press.

Szinicz, L. (2005). History of chemical and biological warfare agents. *Toxicol, 214*(3), 167–181.

Tennison, M., Giordano, J., & Moreno, J. D.. (2017). Security threats versus aggregated truths: Ethical issues in the use of neuroscience and neurotechnology for national security. In J. Illes & S. Hossein (Eds.), *Neuroethics: Defining the Issues in Theory, Practice, and Policy.* Oxford: Oxford University Press.

Tractenberg, R. E., FitzGerald, K. T., & Giordano, J. (2015). Engaging neuroethical issues generated by the use of neurotechnology in national security and defense: Toward process, methods, and paradigm. In J. Giordano (Ed.), *Neurotechnology in National Security and Defense: Practical Considerations, Neuroethical Concerns*, (pp. 259–278). Boca Raton, FL: CRC Press.

Trumpener, U. (1975). The road to Ypres: The beginnings of gas warfare in World War I. *J. Modern History*, *47*(3), 460–480.

Wheelis, M. (2003). 'Non-lethal' chemical weapons: A Faustian bargain. *Issues Sci Technol, 19*(1), 74–78.

Wurzman, R., & Giordano, J. (2015). NEURINT and neuroweapons: Neurotechnologies in national intelligence and defense. In J. Giordano (Ed.), *Neurotechnology in National Security and Defense: Practical Considerations, Neuroethical Concerns* (pp. 79–114). Boca Raton, FL: CRC Press.

ETHICAL CHALLENGES OF INFORMED CONSENT, DECISION-MAKING CAPACITY, AND VULNERABILITY IN CLINICAL DEMENTIA RESEARCH

Pablo Hernández-Marrero, Sandra Martins Pereira, Joana Araújo and Ana Sofia Carvalho

ABSTRACT

This chapter aims to provide an overview of the ethical framework and decision-making in clinical dementia research, and to analyze and discuss the ethical challenges and issues that can arise when conducting clinical dementia research.

Informed consent is the most scrutinized and controversial aspect of clinical research ethics. In clinical dementia research, assessing decision-making capacity may be challenging as the nature and progress of each disease influences decision-making capacity in diverse ways. Persons with dementia represent a vulnerable population deserving special attention when developing, implementing, and evaluating the informed consent process. In this chapter, particular attention will be given to vulnerability categories and how these influence decision-making capacity. Ethical frameworks with a pragmatic contour and implication are needed to protect vulnerable patients from potential harms and ensure their optimal participation in clinical dementia research.

In addition, this chapter analyses important ethical challenges and issues in clinical dementia research. If handled thoughtfully, they would not pose insuperable barriers to research. But if they are ignored, they could slow the

Ethics and Integrity in Health and Life Sciences Research
Advances in Research Ethics and Integrity, Volume 4, 147–168

ISSN: 2398-6018/doi:10.1108/S2398-601820180000004010

research process, alienate potential study subjects and cause harm to research participants. Ethical considerations in research involving persons with dementia primarily concern the representation of the interests of the participants with dementia and protection of their vulnerabilities and rights.

A core set of ethical questions and recommendations are drawn to aid researchers, institutional review boards and potential research participants in the process of participating in clinical dementia research.

Keywords: Research ethics; dementia; dementia research; informed consent; decision-making capacity; vulnerability; autonomy.

INTRODUCTION

Implementing clinical dementia research is marked by several challenges and ethical issues. These may occur in the design, conduct, and monitoring of research and parallel other ethical issues related to involving human subjects as research participants generally (Dunn & Misra, 2009). Due to their vulnerability and cognitive impairment, persons with dementia may be prevented from being included and participating in relevant clinical studies. Nevertheless, in order to ensure evidence-based clinical practices, it is of foremost relevance that these patients are included as research participants. In fact, regardless of the efforts at dementia prevention, important research continues to investigate ways to alleviate clinical dementia's symptoms and improve the quality of life and well-being of these patients, which requires additional human subjects' protections to ethically enroll persons with dementia in clinical dementia research (Johnson & Karlawish, 2015).

In 2011, the International Association of Gerontology and Geriatrics (IAGG) established a global agenda for clinical research in nursing homes. Aligned with this document, other international associations worldwide have been calling for further clinical dementia research, based on the rationale that patients with dementia have the right to take part in clinical and other research studies that can help to produce the evidence base for their care (Rubinstein, Duggan, Landingham, Thompson, & Warburton, 2015; Tolson et al., 2011; World Health Organization, 2015; Wortmann, 2012). According to these associations, failure to include patients with dementia in relevant research is an unacceptable inequality.

Dementia is a terminal condition affecting 6.4 million people across Europe and is recognized by the European Parliament as a major societal challenge. In Europe, the number of people with dementia is predicted to increase from 9.95 million in 2010 to 18.65 million by 2050 (Prince et al., 2013; Tolson et al., 2016; Wortmann, 2012). Given the estimates that rates of dementia will increase to alarming numbers, and no current treatment exists to cure the disease, research to support the needs of this vulnerable population will be critical (Black et al. 2008). In fact, dementia clinical care and research can only progress if individuals with the condition participate in research studies. It is a widely held belief that persons with dementia should be included rather than excluded in studies that would benefit themselves or others (Grout, 2004; Hougham, 2005; Juaristi & Dening, 2016). The question of inclusion and how to best navigate the consent process is part of the ethical and methodological challenge that exists in dementia research (Aselage, Conner, & Carnevale, 2009; Hellström, Nolan, Nordenfelt, & Lundh, 2007).

Excluding vulnerable patients from participating in relevant research would suggest that society is failing in its obligation to improve the standard of health care provided to vulnerable persons, such as those who lack decision-making capacity, by a misguided paternalism. As a result, vulnerable individuals' exclusion from research might condemn them to poor quality care as a result of "evidence biased" medicine (Shepherd, 2016).

Ethics plays a crucial role when integrating medical science, patient choice, and costs in making appropriate decisions. However, it seems that ethical issues are inconsistently addressed in most dementia guidelines, thus it is of foremost relevance to raise awareness and understanding of the complexity of ethical issues in clinical dementia care and research (Knüppel, Mertz, Schmidhuber, Neitzke, & Strech, 2013; Strech, Mertz, Knüppel, Neitzke, & Schmidhuber, 2013; West, Stuckelberger, Pautex, Staaks, & Gysels, 2017). While clinical care and research ethics involving persons with dementia raise identical ethical issues, this chapter will mainly focus and explore these issues when applied to clinical dementia research. Therefore, this chapter aims to provide an overview of the ethical framework and decision-making in clinical dementia research, and to analyze and discuss the ethical challenges and issues that can arise when conducting clinical dementia research.

ETHICAL FRAMEWORK AND DECISION-MAKING IN CLINICAL DEMENTIA RESEARCH

The Ethical Principle of Autonomy

Autonomy is the freedom and ability to act in a self-determined manner. It represents the right of a rational person to express personal decisions independent of outside interference and to have these decisions honored. The principle of autonomy is sometimes described as *respect for* autonomy (Beauchamp & Childress, 2013).

In the domain of health care, respecting a patient's autonomy includes: obtaining informed consent for treatment or for participating in research; facilitating and supporting patients' choices about treatment options; allowing patients to refuse treatments; disclosing comprehensive and truthful information, diagnoses, and treatment options to patients; and maintaining privacy and confidentiality. Respecting autonomy also is important to ensure patients and participants' optimal participation in clinical dementia research. It is also relevant for patients to be able to say no, for treatment or for participation in research, as part of exercising their autonomy.

In the medical literature, an autonomous decision is commonly understood as an act of self-determination exercised by a competent person (Brudney, 2009). In the research context, an autonomous decision requires that research participants have the capacity to provide informed consent to take part in research. Those who cannot make autonomous decisions are either excluded from the study or participate on the basis of a surrogate's consent. Since most policy recommendations limit the types of research that may be conducted with a surrogate's consent (Wendler & Prasad, 2001), it is crucial to know the empirical conditions under which such policies will be implemented (Kim, 2011).

Autonomy and Informed Consent

Informed consent is the most scrutinized and controversial aspect of clinical research ethics. "Respect for persons" provides the foundation for the doctrine of informed consent. The informed consent process comprises three unique components: (1) disclosure of pertinent information including risks, benefits and alternatives; (2) decision-making capacity; and (3) voluntariness (a free and genuine choice made without coercion) (del Carmen & Joffe, 2005; Griffith, 2009; Gupta, 2013; Meisel, Roth, & Lidz, 1997). Hellström et al. (2007) viewed informed consent as a process of consent and assent, which should be monitored throughout the research process. Informed consent must be an ongoing process of communication, understanding and decision-making that involves a wide range of key stakeholders (the patient and potential participant, possible surrogates, clinicians and researchers) throughout the course of the study (Barron, Duffey, Byrd, Campbell, & Ferrucci, 2004).

Informed consent in regard to patients' treatment and research is a legal and ethical issue of autonomy. At the heart of informed consent is respecting a person's autonomy to make personal choices based on the appropriate appraisal of information about the actual or potential circumstances of a situation. Although all conceptions of informed consent must contain the same basic elements, the description of these elements is presented differently by different authors. Beauchamp and Childress (2013) outlined informed consent according to seven elements within three broader categories: (i) *Threshold elements* ((1) competence (to understand and decide) and (2) voluntariness (in deciding)); (ii) *information elements* ((3) disclosure (of material information), (4) recommendation (of a plan) and (5) understanding of both disclosure and recommendation); and (iii) *consent elements* ((6) decision (in favor of a plan) and (7) authorization (of the chosen plan)).

Dempski (2009) presented three basic elements that are necessary for informed consent to take place:

(1) Receipt of information. This includes receiving a description of the procedure, information about the risks and benefits of having or not having the treatment, reasonable alternatives to the treatment, probabilities about outcomes, and "the credentials of the person who will perform the treatment" (Dempski, 2009, p. 78). Since it is too demanding to inform a patient of every possible risk or benefit involved with every treatment or procedure, the obligation is to provide information a reasonable person would want and need to know. Information should be tailored specifically to a person's

personal circumstances, including providing information in the person's spoken language.

(2) Consent for treatment must be voluntary. A person should not be under any influence or be coerced to provide consent. This means that patients should not be asked to sign a consent form when they are under the influence of mind-altering substances, such as narcotics. Depending on the circumstances, consent may be verbalized, written, or implied by behavior. Silence does not convey consent when a reasonable person would normally offer another sign of agreement.

(3) Persons must be competent. Persons must be able to communicate consent and to understand the information provided to them. If a person's condition warrants transferring decision-making authority to a surrogate, informed consent obligations must be met with the surrogate.

Furthermore, Appelbaum and Grisso (1995) stated that the criteria for valid consent to medical treatment should have three elements. The patient must (1) be given adequate information regarding the nature and purpose of proposed treatment, as well as the risks, benefits, and alternatives to the proposed therapy, including no treatment; (2) be free from coercion; and (3) have medical decision-making capacity. The latter element is a key element. Four abilities are commonly used to assess whether or not a patient has medical decision-making capacity. Patients must have the ability to (1) understand the relevant information about proposed diagnostic tests, treatment or research, (2) appreciate their situation (including their underlying values and current medical situation), (3) use reason to make a decision, and (4) communicate their choice (Appelbaum & Grisso, 1988). A fifth legal standard of "reasonableness of choice" can be added to this model (Marson, Ingram, Cody, & Harrell, 1995; Sugarman, McCrory, & Hubal, 1998).

Similarly, Gupta (2013) points out that an informed consent resides on its three critical and essential elements, including voluntarism, information disclosure, and decision-making capacity. In what is referred to as "voluntarism," this element is defined as the ability of an individual to judge, freely, independently, and in the absence of coercion, what is good, right, and best subjected to his/her own situation, values, and prior history (Roberts, 2002). This means that for a consent to be ethically valid, the subject has to make a voluntary decision. Nevertheless, it is known that the voluntarism of an individual may be affected by various factors, namely: intellectual and emotional maturity to make complex decision; illness-related considerations such as the psychological effects of dreaded or incurable diseases or severe mental disorders; religious and cultural values and beliefs; relationships with family or informal caregivers including economic and care burdens; and undue influence or coercion for research participation. The voluntarism of vulnerable subjects is usually compromised; therefore, it is paramount to consider and apply special cautious measures when inviting vulnerable patients for research participation and obtaining their consent.

The second critical element, information disclosure, refers to providing information that is necessary for a patient to make an informed decision and is one of the essential elements of a valid informed consent. The process of information disclosure appears fairly straightforward; however, in real situations it may present difficulties. How much or up to what extent the information should be provided on various aspects of research, such as risks and benefits associated with study intervention, is not clear and is rather a subjective consideration depending on the investigator. Therefore, researchers are recommended to provide the study-related information adequately, judiciously, and truly maintaining an ethical balance between expected risks and benefits of the intervention under investigation.

The last critical element refers to the decision-making capacity, which is defined as the ability to understand and appreciate the nature and consequences of health and/or research participation decisions and to formulate and communicate decisions in this respect. Decisional capacity of an individual depends on his/her cognitive abilities and voluntarism and is adversely affected by cognitive impairment or compromised voluntarism. Decisional capacity comprises four elements or abilities of (1) understanding the information; (2) appreciation of the situation; (3) rational manipulation of the information; and (4) communicating or evidencing a choice (Appelbaum & Grisso, 1988; Fisher, Cea, Davidson, & Fried, 2006). It is worth mentioning that a capable individual must be able to have a factual understanding of the information provided to him/her. Nevertheless, there is much less clarity on what degree or extent of understanding is required for being capable, and on how much information, as a threshold value, must be understood by the individual to be considered as "enough" factual understanding.

Researchers are advised to ensure that the patient has, at least, understood the purpose of the research, risks associated with the research intervention, obligations and consequences of research participation, and his/her right of withdrawing consent any time during the study. For a subject to be considered as being capable of making healthcare decision, he/she must be able to appreciate his/her situation realistically. The research participant must be able to appreciate his/her health condition, any consequences if the condition is left untreated, that the purpose of study is research and not treatment, and the consequences of participation in a research study. To demonstrate decision-making capacity, participants should be able to rationally interpret or adapt the provided or disclosed information for making a rational or logical reason to base their decision or choice upon. Finally, participants must be able to communicate a reasoned choice or decision taken voluntarily. The choice or decision does not necessarily need to be communicated verbally, but participants must be able to express or communicate their choice and preference in some way. Another critical aspect related to choice communication is that the made choice should be sustained over a reasonable time period; however, the patient retains the right of withdrawing the consent any time. Inconsistency or fluctuation of choice over time might reflect impaired decision-making (Gleichgerrcht, Ibáñez, Roca, Torralva, & Manes, 2010; Trachsel, Hermann, & Biller-Andorno, 2015).

It is also worth mentioning that informed consent in patients with dementia is problematic because of the stability and duration of the consent are uncertain, and may be influenced by events emerging in the course of the study (Reyna, Bennett, & Bruera, 2009). Effective strategies to improve the understanding of informed consent information should be considered when designing materials, forms, policies, and procedures for obtaining informed consent. Other than empirical research that has investigated disclosure and understanding of informed consent information, little systematic research has examined other aspects of the informed consent process. This deficit should be rectified to ensure that the rights and interests of patients who participate in research are adequately protected (Sugarman et al. 1998).

Autonomy and Capacity to Consent

Autonomy and consent capacity are two interrelated terms. As the practice of informed consent rests upon the ethical standards and principle of autonomy and self-determination, consent capacity is a fundamental requirement (Fields & Calvert, 2015).

The criteria for valid consent are, normally, based on three elements. The patient must: (1) be given adequate information about the nature and purpose of proposed treatments, as well as the risks, benefits, and alternatives to the proposed therapy, including no treatment; (2) be free from coercion; and (3) have medical decision-making capacity (Sessums, Zembrzuska, & Jackson, 2011). The standards for whether a patient meets this last element also vary, but are generally based on evaluating abilities. The following cognitive criteria for medical decision-making capacity have been proposed and are widely used in clinical research and practice (Appelbaum, 2007): (1) ability to understand relevant information, (2) ability to appreciate the nature of the disorder and the possibility that treatment could be beneficial, (3) ability to reason about the treatment choices, and (4) ability to communicate a choice.

According to Drane (1984), rather than selecting a single standard of competency, a sliding scale is suggested that requires an increasingly more stringent standard as the consequences of the patient's decision embody more risk. The standard of competency to consent to or to refuse treatment or participation in research depends on the degree of danger or risk of harm of the treatment decision and/or research. Since the goal is to promote informed and free participation by competent patients while protecting incompetent patients from the harmful effects of a bad decision, the stringency of the standard of competency at each level is related to the dangerousness of the treatment/research decision.

Many newly diagnosed people with dementia retain adequate mental capacity to make decisions, including whether or not they wish to participate in research. Therefore, careful attempts should be made to involve these persons in clinical dementia research, yet this seldom happens (Higgins, 2013; Holland & Kydd, 2015).

Decision-making Capacity, Capacity, and Competence

Patients are assumed to have capacity to make medical decisions unless proven otherwise (Barton, Mallik, Orr, & Janofsky, 1996). Capacity is a basic requirement for informed consent and is determined based on the process of the patient's decision-making rather than the final decision itself. In most jurisdictions, the patient is required to demonstrate four abilities to have capacity: (1) ability to appreciate the nature of one's situation and the consequences of one's choices, (2) ability to understand the relevant information, (3) ability to reason about the risks and benefits of potential options, and (4) ability to communicate a choice (Appelbaum & Grisso, 1995; Holzer, Gansler, Moczynski, & Folstein, 1997; Sessums et al., 2011). Capacity is influenced by a variety of factors, including situational, psychosocial, medical, psychiatric, and neurological factors (Holzer et al., 1997; Sessums et al., 2011). Accordingly, capacity exists on a continuum, can be evanescent, and can be optimized (Grisso & Appelbaum, 1998).

Capacity or competence, a related notion, should not be confused with decisional capacity; however they describe an individual's ability to make a decision. In the literature, these two terms – capacity and competence – are commonly used interchangeably. Yet, as Appelbaum (2007) points out, competence refers to legal judgments and capacity to clinical ones. This is also supported by other authors who define competence as an individual's legal standing to make healthcare decisions or as a legal determination made by a court of law (del Carmen & Joffe, 2005; Ganzini, Volicer, Nelson, Fox, & Derse, 2005).

As dementia progresses, persons demonstrate increasing loss of capacity, decision-making capacity, and competence over time (Aselage et al., 2009). While these are intertwined terms, in reality they are defined differently. On one hand, decision-making capacity is generally held to be present when one possesses the ability to understand relevant information, appreciate the nature of the situation and its consequences, reason by manipulating information, and express a choice (Black et al., 2008). On the other hand, capacity differs from decision-making capacity in that the person must be able to express understanding of consequences. Capacity can diminish over time, in that a person can possess capacity in the earlier stages of dementia but during longitudinal research may lose capacity (Dewing, 2007). Finally, competence is linked with the concept of self-awareness, and the problem with assessing cognitively impaired individuals is that cognitive function may be sporadic, fluctuating, and context-dependent (Gleichgerrcht et al., 2010; Trachsel et al., 2015). Therefore, competence is a legal term, and a person's level of competence is assessed by court proceeding to determine a declaration of competence or incompetence (Woods & Pratt, 2005).

Because a patient's capacity is both temporal and situational, capacity evaluations should occur in the context of the specific healthcare decision that needs to be made. Some patients lack capacity for specific periods of time, such as when critically ill, but not permanently. Although some people are completely incapacitated, many have limited capacity. Those with limited capacity may be able to make some diagnostic and treatment decisions (generally less risky decisions) but not others (Sessums et al., 2011).

When there are concerns about a patient's capacity, this should result in a formal assessment. The gold standard for capacity determination is a clinical examination by a physician trained to do the examination and who has performed an extensive number of capacity evaluations. However, several authors have been emphasizing that there is currently no consensus on how to reliably assess decision-making capacity in dementia research participants (Dunn, Nowrangi, Palmer, Jeste, & Saks, 2006; Guarino et al., 2016; Howe, 2012; Karlawish, 2008).

In clinical dementia research, assessing decision-making capacity may be challenging as the nature and progress of each disease influence decision-making capacity in diverse ways. Research on the decision-making capacity of patients with Alzheimer disease highlights the practical challenges for policies that rely on drawing a clear line between capacity and incapacity. This challenge is especially important for studies that attempt to enroll only competent individuals because of substantial research risks (Kim, 2011)

According to Ganzini et al. (2005), there are ten myths that clinicians hold about decision-making capacity: (1) decision-making capacity and competency are the same; (2) lack of decision-making capacity can be presumed when patients go against medical advice; (3) there is no need to assess decision-making capacity unless patients go against medical advice; (4) decision-making capacity is an "all or nothing" phenomenon; (5) cognitive impairment equals lack of decision-making capacity; (6) lack of decision-making capacity is a permanent condition; (7) patients who have not been given relevant and consistent information about their treatment lack decision-making capacity; (8) all patients with certain psychiatric disorders lack decision-making capacity; (9) patients who are involuntarily committed lack decision-making capacity; and (10) only mental health experts can assess decision-making capacity. As patients with dementia are at risk of being considered as lacking capacity to make decisions on their willingness to participate in clinical research, it is of foremost relevance to foster clinicians and researchers' understanding on decision-making capacity in an attempt to prevent potential errors that may be caused by these myths.

The Ethical Principle of Vulnerability

Vulnerability is a touchstone in bioethics and a common denominator to both clinical and research ethics. As a concept, vulnerability is defined as the susceptibility of being wounded or hurt, being a term of Latin origin, meaning *vulnus*, wound. This concept was introduced into the vocabulary of bioethics in the sphere of human experimentation, as a characteristic attributed to particular populations considered to be those most exposed to and poorly defended against maltreatment or abuse of others. Substantial debates have been established to define who vulnerable people are and to minimize the risk they have of being exploited. Defining vulnerable persons or populations has proved to be more difficult than initially expected (Hurst, 2008).

Kipnis (2001) was one of the first authors suggesting that vulnerability is inherent in situations and not persons, thus challenging national and international guidelines for research that identify vulnerability as specific classes of people (e.g., persons with dementia). According to Kipnis (2001), specific categories of vulnerability can be identified as opposed to the institutionalized subpopulation focus (e.g., children, prisoners, pregnant women, persons with dementia or cognitive impairment). Researchers should take this into account. The following distinct characteristics of vulnerability were identified by Kipnis (2001) who also indicated how to identify each one of them through a single question for each characteristic:

(1) Cognitive vulnerability: does the person have the capacity to understand and decide to be treated or not?

    This category includes, for instance, persons with early-stage dementia, with certain types of mental illness, intellectual disability, older adults who are institutionalized in nursing homes, older persons with educational deficits and/or unfamiliarity with the medical language. Indeed, persons with dementia represent a vulnerable group of individuals who deserve special attention when developing, implementing and evaluating the informed consent process.

(2) Situational vulnerability: is the patient in a situation in which medical exigency prevents the education and deliberation needed to decide?

    This category includes patients who cannot be sufficiently informed and/or who cannot complete effective deliberation within the available timeframe. A few examples can be given on this situation: patients receiving a very serious diagnostic or closed prognostic (e.g., dementia), patients who are being submitted to an urgent surgery, or even patients who need to be sedated for a certain reversible reason.

(3) Medical vulnerability: has the person a serious health-related condition for which there is no satisfactory solution?

    Some patients could be more vulnerable due to a set of different circumstances. Therefore, a patient can have additional vulnerability when he/she has a serious health-related condition for which there are no satisfactory treatments. Once again, patients with dementia can experience this type of vulnerability. Other examples include, for instance, patients with palliative care needs, with metastatic cancers, Parkinson's disease, multiple sclerosis, etc.

(4) Allocational vulnerability: is the patient seriously lacking in important social goods that will influence his/her decision?

    Very poor patients and all the older patients who are in a socially devaluated or disadvantaged condition can be identified as having this type of vulnerability.

(5) Social vulnerability: does the patient belong to a group whose rights and interests have been socially disvalued?

    As an example of social vulnerability we can include persons with addictions, older patients, older migrants, homeless, prisoners and all those persons who might be in a socially disvalued or disadvantaged situation.

(6) Deferential vulnerability: is the patient deferential behavior what masks an underlying unwillingness to decide?

> This type of vulnerability refers to powerful social and cultural pressures. It is known, for example, that some patients experience the "white coat syndrome" toward their medical doctors, or, in another scope, women or men who may find it hard to turn down requests from their spouses.

In fact, aging, in combination with other factors, increases vulnerability and puts one at risk for decision-making impairments. Combined vulnerabilities may occur more frequently among older people or patients with dementia. This is more likely to have an impact on decision-making than any individual vulnerability (Christensen, Haroun, Schneiderman, & Jeste, 1995). Overall, future research efforts should allow us to develop a more accurate understanding and identification about how vulnerability (and vulnerabilities) may influence decision-making capacity and the development of more effective instruments that will enable the participation of vulnerable persons with dementia in clinical care and research decision-making process.

If we can more accurately recognize vulnerable individuals, we will ultimately be better at protecting them from improper consent, avoiding cases where too much protection results in lack of access of appropriate treatment or research participation (Cherubini et al., 2011). It is worth mentioning that informed consent is only ethically valid or legitimate if we take vulnerabilities into consideration when assessing decision-making capacity. Furthermore, adequate and creative strategies need to be designed to empower those with vulnerabilities. This is a major challenge for researchers, physicians and ethicists.

Persons with dementia are vulnerable for various reasons: they are often dependent on others to meet their care needs; they are at increased risk of adverse effects associated with experimental treatments; they are often prevented of accessing proper care, particularly at the end of life; and their ability to make an informed choice about both what type of care they would like to receive and participating in research may be reduced temporarily, intermittently or permanently. This vulnerability calls for protection, but it does not mean that participation in research should be prohibited. On the contrary, vulnerable persons, including patients with dementia, may and should be included in research if research bears direct relevance to their medical condition and carries the opportunity of medical benefit to those enrolled, or the opportunity to advance knowledge in a specific field or for a specific group of patients (Bruera, 1994; Reyna et al., 2009; Welch et al., 2015). Recognizing the risk of enhanced vulnerability, specific safeguards need to be put in place to protect and prevent these participants from any harm. Rather than checking whether a specific group of patients is vulnerable (e.g., patients with dementia), efforts need to be made to carefully assess whether a potential research participant is vulnerable or not based on the risk of harm research participation entails for him/her. Ethical frameworks with a pragmatic contour and implication are needed to protect vulnerable patients from potential harms and ensure their optimal participation in clinical dementia research (Wendler, 2017).

ETHICAL CHALLENGES AND ISSUES IN CLINICAL DEMENTIA RESEARCH

Dementia is highly prevalent, incurable, and devastating. Research is, therefore, crucial but involves patients with substantial cognitive impairment that may hinder their ability to provide informed consent (Kim, Caine, Currier, Leibovici, & Ryan, 2001; Okonkwo et al., 2007). The problem is magnified when the research is invasive and burdensome with unpredictable risks (Orgogozo et al., 2003).

Despite several decades of worldwide debates, no clear policy exists regarding the involvement of adults with decisional impairments in clinical research (Saks, Dunn, Wimer, Gonzales, & Kim, 2008). In fact, although there has been much standardization of the ethical requirements to undertake clinical dementia research, culturally different approaches still remain throughout the world. Patients with dementia represent thus a highly vulnerable group.

Although policies on the ethics of dementia research remain unresolved, the emerging empirical literature, however, shows two important trends that should inform future policy discussions. First, research on the decision-making capacity of patients with dementia highlights the practical challenges for policies that rely on drawing a clear line between capacity and incapacity. This challenge is especially important for studies that attempt to enroll only competent individuals because of substantial research risks. Second, evidence is accumulating that incapacitated patients with dementia can be enrolled in research in ways that are consistent with their values, given that important, ethically relevant abilities may be preserved. Research also shows consistent and broad support for surrogate consent, advanced consent and supported decision-making for dementia research (even for protocols with invasive procedures) among various lay stakeholder groups.

According to Emanuel, Wendler, and Grady (2000), research is ethically sound when it meets seven major requirements: (1) *value*–enhancements of health or knowledge; (2) *scientific validity*–ensuring that the research is methodologically sound; (3) *fair participant selection* – scientific objectives, not vulnerability or privilege, and the potential for and distribution of risks and benefits, should determine communities selected as study sites and the inclusion criteria for individual participants; (4) *favorable risk-benefit ratio* – within the context of standard clinical practice and the research protocol, risks must be minimized, potential benefits enhanced, and the potential benefits to individuals and knowledge gained for society must outweigh the risks; (5) *independent review* – unaffiliated individuals must review the research and approve, amend, or terminate it; (6) *informed consent* – individuals should be informed about the research and provide their voluntary consent; and (7) *respect for enrolled subjects* – subjects should have their privacy protected, the opportunity to withdraw, and their well-being monitored. Fulfilling all these seven requirements is necessary and sufficient to make clinical research ethical.

Ethical challenges in clinical dementia research are multiple and include communication challenges, trustworthiness of data when including cognitive impaired patients as participants, respecting participants' abilities to make decisions, securing consent to participate from proxies or surrogate decision-makers, or simply excluding persons with dementia from research because of their cognitive impairment and vulnerabilities (Beuscher & Grando, 2009; Hellström et al., 2007; Slaughter, Cole, Jennings, & Reimer, 2007). Beyond these, other ethical issues may occur intimately linked to the research process itself, namely: (1) the necessity of very large cohorts of research subjects, recruited for lengthy studies, probably ending only in participants' deaths; (2) the creation of cohorts of "study ready" volunteers, many of whom will be competent to consent at the beginning of the process, but move into cognitive impairment later; (3) reliance on adaptive trial design, creating challenges for informed consent, equipoise and justice; (4) the use of biomarkers and predictive tests that describe risk rather than certainty, and that can threaten participants' welfare if the information is obtained by insurance companies or long-term care providers; (5) the use of study partners that creates unique risks of harm to the relationship of subject and study partner (Davis, 2017). Greater attention is needed, at all levels, to face and overcome these complex ethical issues.

Informed Consent in Persons Without Capacity to Consent

Cognitive impairments in patients with dementia, together with other disorders affecting cognition, may have a negative impact on their capacity to provide consent to clinical care or research participation (Appelbaum, 2010). In addition, ageism among investigators can also contribute to the failure of informed consent (Barron et al., 2004) and prevent patients with dementia from being included as participants in relevant clinical research.

Adults with cognitive impairment are considered a vulnerable population without capacity to consent. The conditions associated with cognitive impairment, such as dementia and delirium, cause great suffering to affected patients and their families. Improving clinical care for these conditions depends on research involving cognitively impaired participants. Cognitive impairment is at times associated with partial or full impairment of the capacity to consent to research. This both limits the ability of the individual to consent personally to research participation, and also increases pressure upon Institutional Review Boards (IRBs)/Research Ethics Committees (RECs) and investigators to place additional safeguards for the appropriate participation of cognitively impaired individuals in research.

Special protection is to be given to persons who do not have the capacity to consent. In these cases, research should only be carried out for the benefit of the participant, subject to the authorization and protective condition prescribed by law, and if there is no research alternative of comparable effectiveness with participants able to consent (Martin, 2009).

While the ethical and legal principles permitting and safeguarding the participation of cognitively impaired persons in research are generally agreed upon, there are no specific methods that operationalize these principles in a language that can be used by Institutional Review Boards and researchers to guide their day-to-day work in this area. In situations in which IRBs/RECs might not have specific policies in this area, gold-standard ethical guidelines may also serve as the foundation for such policies. In fact, as aforementioned, cognitively impaired participants might be involved in a research project. For this to happen, researchers need to consider the use of screening for cognitive impairment, the conduct of assessments evaluating capacity to consent to research, situations in which proxies might consent for research participation in the place of cognitively impaired participants, how to go about identifying appropriate proxies, and how to deal with the loss of consent capacity in the course of a research project (Alzheimer's Association, 2004).

Persons with impaired decision-making capacity require special attention and ethical considerations and procedures during recruitment and participation in research. When substitute consent is necessary, state laws generally provide a range of options, including proxy consent and advance directives.

Proxy Consent

When the patient is not capable of making a decision and consenting, consent can (and should) be obtained by proxy. In the literature, a proxy is also named as a legal authorized representative, surrogate decision-maker, or substitute decision-maker, and this figure can be a family member, close friend, or legally appointed guardian (Aselage et al., 2009; Dewing, 2007; Dunn et al., 2013; Kim et al., 2009; Silverman, Luce, & Schwartz, 2004; Slaughter et al., 2007). Proxy (or substitute or surrogate) decision-making is the situation in which decisions are made by a proxy when the research participant is unable to consent. The proxy plays the role that the research participant would play if he or she were competent and capable, and makes decisions on behalf (and for the benefit) of the research participant. Three general standards have been proposed for this type of approach: subjective standard, substituted judgment, and best interest (Black, Wechsler, & Fogarty, 2013; Kluge, 2008).

Practices of proxy consent however vary largely across countries. While in some cases, authorized representatives may be legally mandated by individuals, for instance, via an advanced directive, in other cases family members are recognized by professional staff and researchers as sharing responsibility for family care and participation in research. Despite country variations, the general consensus is that proxies should make decisions about research participation in a similar manner as they make decisions about clinical care (Karlawish, 2003; Karlawish, Casarett, Klocinski, & Sankar, 2001; Reyna et al., 2009). In fact, a proxy for research decisions would typically be the same person chosen by the patient to make clinical decisions. The requirement to add surrogate consent becomes more stringent as the risk of participating in the study increases and the potential benefit for the participant decreases (Barron et al., 2004). The role and involvement of carers in the decision-making process needs to be considered so that potential participants are not unnecessarily excluded (Agarwal, Ferran, Ost, & Wilson, 1996).

With regard to proxy consent, a number of issues require further study. These include how state laws address (or fail to address) research involving cognitively impaired individuals and what effects this has on research conduct (Appelbaum, 2002; Kim, Appelbaum, Jeste, & Olin, 2004); how IRBs/RECS define and weigh risks and benefits in considering research involving proxy consent (Silverman, Hull, & Sugarman, 2001); how various stakeholders, including the general public, people with disorders that may impair decision-making capacity, and proxies themselves view proxy consent for research (Kim et al., 2009); and to what degree proxies' research decisions reflect what patients themselves would decide (Dunn & Misra, 2009). This last issue requires particular attention for researchers. In fact, when engaging proxy decision-makers in the informed consent process, researchers should instruct them to make decisions based on the persons with dementia's expressed wishes, values and beliefs (Cacchione, 2011).

Advanced Consent

Another way of obtaining consent for research participation in persons with dementia is through advanced consent; the so-called "research advance directive" (Barron et al., 2004; Jongsma & van de Vathorst, 2015a, b; Porteri, 2018; Reyna et al., 2009). Research advance directives (also named advance research directives) have been proposed as a mechanism for prospective consent for persons who anticipate cognitive impairment, as in the case of prodromal or early-stage dementia patients (Pierce, 2010).

Under certain conditions, this might be the best available option to assess whether or not a person with dementia consents to research participation. The practical utility of an advance directive for clinical dementia research is even higher than for medical treatment as the balance between benefit and risk is more difficult to estimate (Porteri, 2018). In addition, there is some evidence suggesting that a directive might provide better insight into a person's wishes than the person's proxy, although neither source is perfect (Bravo, Sene, & Arcand, 2018).

Nevertheless, several authors have emphasized that this type of instrument has been receiving minimal support for their expansion into the context of research, particularly due to the need to further assess whether participation falls under the umbrella of "clinical interest" (Buller, 2015; Jongsma & van de Vathorst, 2015a, b). Moreover, the existence of an advance directive or advanced consent for research participants poses particular challenges. While this type of document is usually legally binding, there are still some doubts and insecurities about the possibility of revocation (Schmidhuber, Haeupler, Marinova-Schmidt, Frewer, & Kolominsky-Rabas, 2017). The truth is that there are no general answers to properly address this issue, which highlights the need to further study the use of advanced consent in clinical dementia research.

Supported Decision-making

Supported decision-making refers to scenarios where decisions are made by both the person with dementia and another person or persons who provide different levels and types of support. The core ethical principle underlying supported decision-making is autonomy. In other words, supported decision-making means that no potential research participants should have somebody else appointed to make the decision on their behalf, if they could make the decision themselves with assistance and support (Black et al., 2013; Davidson et al., 2015; Keeling, 2016).

In the context of clinical dementia research, supported decision-making offers the possibility to actually involve the person with dementia in the decision-making process. At the same time, this emphasizes the relational nature of autonomy and highlights beneficence and solicitude as other ethical principles to be considered. According to Scholten and Gather (2018), a combined model integrating the strengths of the competence model with supported decision-making may constitute and alternative way forward. Further empirical research is needed to clarify who may benefit most from supported decision-making, decisions in need of support, selection of supporters, guidelines for the supported decision-making process, integration of supported decision-making with emerging technological platforms, and outcomes of supported decision-making (Jeste et al., 2018).

CONCLUSION

This chapter provides an overview of an ethical framework and ethical principles linked to informed consent, decision-making capacity and vulnerability in clinical dementia research. In addition, this chapter analyses important ethical challenges and issues in this area. If handled thoughtfully, these ethical challenges and issues will not pose insuperable barriers to research. But if they are ignored, they can slow the research process, alienate potential study participants, and cause harm to research participants. Ethical considerations in research involving persons with dementia primarily concern the representation of the interests of the participants with dementia and protection of their vulnerabilities and rights.

Although people with dementia were once commonly excluded from research (Hubbard, Downs, & Tester, 2003), researchers increasingly are encouraged to involve them as participants in research on care and services. Including people with dementia as research participants is particularly important because proxy participants, such as family members or caregivers, often have different views and provide different data than people with dementia. Ethical principles that guide researchers, including respect for persons, beneficence, and justice, all support and help guide the involvement of people with dementia in research.

A core set of ethical questions and recommendations are drawn as follows to aid researchers, institutional review boards and potential research participants in the process of participating in clinical dementia research.

3.1. Ethical Questions

What is the likelihood that people with dementia, such as Alzheimer disease, will be competent to give informed consent?

Does the person have the capacity to understand and decide to be treated or not?

Is the patient in a situation in which medical exigency prevents the education and deliberation needed to decide?

Has the person a serious health-related condition for which there is no satisfactory solution?

Is the patient seriously lacking in important social goods that will influence his/her decision?

Does the patient belong to a group whose rights and interests have been socially disvalued?

Is the patient's deferential behavior possibly masking an underlying unwillingness to decide or a perceived pressure to participate?

Is the exclusion from research a harm to vulnerable people as they may be denied access to understanding how health interventions work for them in clinical settings?

Are any potential research participants at risk of being harmed in any way by being included or excluded from participating in this research study?

In case of advanced consent via an advance research directive, how and in under what conditions should a revocation be accepted?

Which expressions (verbal, mimic or gesticulatory) should be taken into account as expressions of an autonomous will?

How best can diverse models of decision-making be integrated in clinical dementia research?

3.2. Recommendations

Researchers should be aware that persons with dementia are individuals and have value to offer to the research arena.

Researchers should consider persons with dementia not only as subjects and research participants, but as partners in the research process.

A universal ethical approach and policy recommendations to obtaining informed consent and monitoring the appropriateness of research should be developed.

Informed consent processes and practices should be conducted using a simple language, having a concise format, giving patients and/or relatives/legal guardians enough time to read the consent form carefully and discuss it with other family member(s) or professional(s).

Creative strategies can be used to optimize comprehension among frail persons with dementia in multiples formats and media (e.g., simplified storybooks, videos, vignettes).

Informed consent is a process and might require more than one short session for patients to absorb and understand the research issues.

A multifaceted decision-making model, integrating different sources of information, might better serve the interests of persons with dementia who have lost the capacity to make decisions on their own.

Guidelines on ethics procedures in clinical dementia research are needed to address ethical issues and how to deal with them. This will help health professionals to better understand how to approach patients with dementia, and for patients, their relatives, and the general public seeking for information and advice.

Because ethics procedures in clinical dementia research are commonly very brief and focused on the reference of having obtained ethics approvals, journals ought to include and request authors to provide more detailed information on ethics procedures in their guidelines for submissions.

ACKNOWLEDGMENT

This chapter was written during the duration of Subproject ETHICS II of Project ENSURE "Enhancing the Informed Consent Process: Supported Decision-Making and Capacity Assessment in Clinical Dementia Research." The authors would like to thank ERA-NET NEURON II, ELSA 2015, European Commission, and Fundação para a Ciência e a Tecnologia (FCT), Ministério da Ciência, Tecnologia e Ensino Superior, Portugal, for their financial support to the Subproject ETHICS II of Project ENSURE. The funders had no role in the design and writing of this chapter.

REFERENCES

Agarwal, M. R., Ferran, J., Ost, K., & Wilson, K. C. (1996). Ethics of 'informed consent' in dementia research – the debate continues. *International Journal of Geriatric Psychiatry*, *11*(9), 801–806.

Alzheimer's Association. (2004). Research consent for cognitively impaired adults: recommendations for institutional review boards and investigators. *Alzheimer Disease & Associated Disorders*, *18*(3), 171–175.

Appelbaum, P. S. (2002). Involving decisionally impaired subjects in research: The need for legislation. *American Journal of Geriatric Psychiatry*, *10*(2), 120–124. doi:10.1097/00019442-200203000-00002.

Appelbaum, P. S. (2007). Clinical practice. Assessment of patients' competence to consent to treatment. *New England Journal of Medicine*, *357*(18), 1834–1840. doi:10.1056/NEJMcp074045.

Appelbaum, P. S. (2010). Consent in impaired populations. *Current Neurology and Neuroscience Reports*, *10*(5), 367–373. doi:10.1007/s11910-010-0123-5.

Appelbaum, P. S., & Grisso, T. (1988). Assessing patients' capacities to consent to treatment. *New England Journal of Medicine*, *319*(25), 1635–1638. doi:10.1056/NEJM198812223192504

Appelbaum, P. S., & Grisso, T. (1995). The MacArthur treatment competence study: I, mental illness and competence to consent to treatment. *Law and Human Behaviour*, *19*(2), 105–126.

Aselage, M., Conner, B., & Carnevale, T. (2009). Ethical issues in conducting research with persons with dementia. *Southern Online Journal of Nursing Research*, *9*(4), 1–21.

Barron, J. S., Duffey, P. L., Byrd, L. J., Campbell, R., & Ferrucci, L. (2004). Informed consent for research participation in frail older persons. *Aging Clinical Experiment Research*, *16*(1), 79–85.

Barton, C. D. Jr., Mallik, H. S., Orr, W. B., & Janofsky, J. S. (1996). Clinicians' judgement of capacity of nursing home patients to give informed consent. *Psychiatric Services*, *47*(9), 956–960. doi:10.1176/ps.47.9.956

Beauchamp, T. L., & Childress, J. F. (2013). *Principles of biomedical ethics*. (7th edition). Oxford: Oxford University Press.

Beuscher, L., & Grando, V. T. (2009). Challenges in conducting qualitative research with persons with dementia. *Research in Gerontological Nursing*, *2*(1), 6–11. doi:10.3928/19404921-2009 0101-04.

Black, B., Brandt, J., Rabins, P., Samus, Q. M., Steele, C. D., Lyketsos, C. G., & Rosenblatt, A. (2008). Predictors of providing informed consent or assent for research participation in assisted living residents. *American Journal of Psychiatry*, *16*(1), 83–91.

Black, B. S., Wechsler, M., & Fogarty, L. (2013). Decision making for participation in dementia research. *American Journal of Geriatric Psychiatry*, *21*(4), 355–363. doi:10.1016/j.jagp.2012.11.009.

Bravo, G., Sene, M., & Arcand, M. (2018). Making medical decisions for an incompetent older adult when both a proxy and an advance directive are available: Which is more likely to reflect the older adult's preferences? *Journal of Medical Ethics*, *44*(7), 498–503. doi:10.1136/medethics-2017-104203.

Brudney, D. (2009). Choosing for another: Beyond autonomy and best interests. *Hastings Center Report*, *39*(2), 31–37.

Bruera, E. (1994). Ethical issues in palliative care research. *Journal of Palliative Care*, *10*(3), 7–9.

Buller, T. (2015). Advance consent, critical interests and dementia research. *Journal of Medical Ethics*, *41*(8), 701–707. doi:10.1136/medethics-2014-102024.

Cacchione, P. Z. (2011). People with dementia: Capacity to consent to research participation. *Clinical Nursing Research*, *20*(3), 223–227.

Cherubini, A., Oristrell, J., Pla, X., Ruggiero, C., Ferretti, R., Diestre, G., … Mills, G. H. (2011). The persistent exclusion of older patients from ongoing clinical trials regarding heart failure. *Archives of Internal Medicine*, *171*(6), 550–556. doi:10.1001/archinternmed.2011.31.

Christensen, K., Haroun, A., Schneiderman, L. J., & Jeste, D. V. (1995). Decision-making capacity for informed consent in the older population. *Journal of the American Academy of Psychiatry and the Law Online*, *23*(3), 353–365.

Davidson, G., Kelly, B., Macdonald, G., Rizzo, M., Lombard, L., Abogunrin, O. … Martin, A. (2015). Supported decision making: A review of the international literature. *International Journal of Law and Psychiatry*, *38*, 61–67. doi:10.1016/j.ijlp.2015.01.008

Davis, D. S. (2017). Ethical issues in Alzheimer's disease research involving human subjects. *Journal of Medical Ethics*, *43*(12), 852–856. doi:10.1136/medethics-2016-103392

del Carmen, M. G., & Joffe, S. (2005). Informed consent for medical treatment and research: A review. *Oncologist*, *10*(8), 636–641. doi:10.1634/theoncologist.10-8-636

Dempski, K. (2009). Informed consent. In: S. J. Westrick & K. Dempski (Eds.), *Essentials of nursing law and ethics*. Boston, MA: Jones & Bartlett.

Dewing, J. (2007). Participatory research: A method for process consent with persons who have dementia. *Dementia*, *6*(1), 11–15. doi:10.1177/1471301207075625

Drane, J. F. (1984). Competency to give an informed consent. A model for making clinical assessments. *Journal of the American Medical Association*, *252*(7), 925–927. doi:10.1001/jama.1984.03350070043021.

Dunn, L. B., Fisher, S. R., Hantke, M., Appelbaum, P. S., Dohan, D., Young, J. P., & Roberts, L. W. (2013). "Thinking about it for somebody else": Alzheimer's disease research and proxy decision makers' translation of ethical principles into practice. *American Journal of Geriatric Psychiatry*, *21*(4), 337–345. doi:10.1016/j.jagp.2012.11.014.

Dunn, L. B., & Misra, S. (2009). Research ethics issues in geriatric psychiatry. *Psychiatric Clinics of North America*, *32*(2), 395–411. doi:10.1016/j.psc.2009.03.007.

Dunn, L. B., Nowrangi, M. A., Palmer, B. W., Jeste, D. V., & Saks, E. R. (2006). Assessing decisional capacity for clinical research or treatment: a review of instruments. *American Journal of Psychiatry*, *163*(8), 1323–1334. doi:10.1176/ajp.2006.163.8.1323.

Emanuel, E. J., Wendler, D., & Grady, C. (2000). What makes clinical research ethical? *Journal of the American Medical Association*, *283*(20), 2701–2711. doi:10.1001/jama.283.20.2701.

Fields, L. M., & Calvert, J. D. (2015). Informed consent procedures with cognitively impaired patients: A review of ethics and best practices. *Psychiatry and Clinical Neurosciences*, *69*(8), 462–471. doi:10.1111/pcn.12289.

Fisher, C. B., Cea, C. D., Davidson, P. W., & Fried, A. L. (2006). Capacity of persons with mental retardation to consent to participate in randomized clinical trials. *American Journal of Psychiatry*, *163*(10), 1813–1820.

Ganzini, L., Volicer, L., Nelson, W. A., Fox, E., & Derse, A. R. (2005). Ten myths about decision-making capacity. *Journal of the American Medical Directors Association*, *6*(3 Suppl), S100–S104.

Gleichgerrcht, E., Ibáñez, A., Roca, M., Torralva, T., & Manes, F. (2010). Decision-making cognition in neurodegenerative diseases. *Nature Reviews Neurology*, *6*(11), 611–623. doi:10.1038/nrneurol.2010.148.

Griffith, R. (2009). Elements of a valid consent to treatment in capable adults. *Journal of Paramedic Practice*, *1*(5), 196–203.

Grisso, T., & Appelbaum, P. S. (1998). *Assessing competence to consent to treatment*. New York, NY: Oxford University Press.

Grout, G. (2004). Using negotiated consent in research and practice. *Nursing Older People*, *16*(4), 18–20.

Guarino, P. D., Vertrees, J. E., Asthana, S., Sano, M., Llorente, M. D., Pallaki, M., … Dysken, M. W. (2016). Measuring informed consent capacity in an Alzheimer's disease clinical trial. *Alzheimers's and Dementia: Translational Research & Clinical Interventions*, *2*(4), 258–266. doi:10.1016/j.trci.2016.09.001.

Gupta, U. C. (2013). Informed consent in clinical research: Revisiting few concepts and areas. *Perspectives in Clinical Research*, *4*(1), 26–32. doi:10.4103/2229-3485.106373.

Hellström, I., Nolan, M., Nordenfelt, L., & Lundh, U. (2007). Ethical and methodological issues in interviewing persons with dementia. *Nursing Ethics*, *14*(5), 608–619. doi:10.1177/0969733007080206.

Higgins, P. (2013). Involving people with dementia in research. *Nursing Times*, *109*(8), 20–23.

Holland, S., & Kydd, A. (2015). Ethical issues when involving people newly diagnosed with dementia in research. *Nurse Research*, *22*(4), 25–29. doi:10.7748/nr.22.4.25.e1308.

Holzer, J. C., Gansler, D. A., Moczynski, N. P., & Folstein, M. F. (1997). Cognitive functions in the informed consent evaluation process: A pilot study. *Journal of the American Academy of Psychiatry and the Law*, *25*(4), 531–540.

Hougham, G. W. (2005). Waste not, want not: Cognitive impairment should not preclude research participation. *American Journal of Bioethics*, *5*(1), 36–37. doi:10.1080/15265160590927732.

Howe, E. (2012). Informed consent, participation in research, and the Alzheimer's patient. *Innovations in Clinical Neuroscience*, *9*(5–6), 47–51.

Hubbard, G., Downs, M. G., & Tester, S. (2003). Including older people with dementia in research: Challenges and strategies. *Aging & Mental Health*, *7*(5), 351–362. doi:10.1080/1360786031000150685.

Hurst, S. A. (2008). Vulnerability in research and health care; describing the elephant in the room? *Bioethics*, *22*(4), 191–202. doi:10.1111/j.1467-8519.2008.00631.x.

Jeste, D. V., Eglit, G. M. L., Palmer, B. W., Martinis, J. G., Blanck, P., & Saks, E. R. (2018). Supported decision making in serious mental illness. *Psychiatry*, *81*(1), 28–40. doi:10.1080/00332747.2017.1324697.

Johnson, R. A., & Karlawish, J. (2015). A review of ethical issues in dementia. *International Psychogeriatrics*, *27*(10), 1635–1647. doi:10.1017/S1041610215000848.

Jongsma, K. R., & van de Vathorst, S. (2015a). Beyond competence: advance directives in dementia research. *Monash Bioethics Review*, *33*(2–3), 167–180. doi:10.1007/s40592-015-0034-y.

Jongsma, K. R., & van de Vathorst, S. (2015b). Dementia research and advance consent: it is not about critical interests. *Journal of Medical Ethics*, *41*(8), 708–709. doi:10.1136/medethics-2014-102445.

Juaristi, G. E., & Dening, K. H. (2016). Promoting participation of people with dementia in research. *Nursing Standard*, *30*(39), 38–43. doi:10.7748/ns.30.39.38.s43.

Karlawish, J. (2008). Measuring decision-making capacity in cognitively impaired individuals. *Neurosignals*, *16*(1), 91–98. doi:10.1159/000109763.

Karlawish, J. H. (2003). Conducting research that involves subjects at the end of life who are unable to give consent. *Journal of Pain and Symptom Management*, *25*(4), S14–S24.

Karlawish, J. H., Casarett, D., Klocinski, J., & Sankar, P. (2001). How do AD patients and their caregivers decide whether to enroll in a clinical trial? *Neurology*, *56*(6), 789–792.

Keeling, A. (2016). Supported decision making: the rights of people with dementia. *Nursing Standard*, *30*(30), 38–44. doi:10.7748/ns.30.30.38.s45.

Kim, S., Kim, H., Langa, K., Karlawish, D., Knopman, D., & Appelbaum, P. (2009). Surrogate consent for dementia research. *Neurology*, *72*(2), 149–155. doi:10.1212/01.wnl.0000339039.18931.a2.

Kim, S. Y., Appelbaum, P. S., Jeste, D. V., & Olin, J. T. (2004). Proxy and surrogate consent in geriatric neuropsychiatric research: Update and recommendations. *American Journal of Psychiatry*, *161*(5), 797–806. doi:10.1176/appi.ajp.161.5.797.

Kim, S. Y. H. (2011). The ethics of informed consent in Alzheimer disease research. *Nature Reviews Neurology*, *7*, 410–414. doi:10.1038/nrneurol.2011.76.

Kim, S. Y. H., Caine, E. D., Currier, G. W., Leibovici, A., & Ryan, J. M. (2001). Assessing the competence of persons with Alzheimer's disease in providing informed consent for participation in research. *American Journal of Psychiatry*, *158*, 712–717. doi:10.1176/appi.ajp.158.5.712.

Kipnis, K. (2001). Vulnerability in research subjects: A bioethical taxonomy. *Ethical and policy issues in research involving human participants*, *2*.

Kluge, E.-H. W. (2008). Incompetent patients, substitute decision making, and quality of life: Some ethical considerations. *The Medscape Journal of Medicine*, *10*(10), 237.

Knüppel, H., Mertz, M., Schmidhuber, M., Neitzke, G., & Strech, D. (2013). Inclusion of ethical issues in dementia guidelines: A thematic text analysis. *PLoS Medicine*, *10*(8), e1001498. doi:10.1371/journal.pmed.1001498.

Marson, D. C., Ingram, K. K., Cody, H. A., & Harrell, L. E. (1995). Assessing the competency of patients with Alzheimer's disease under different legal standards. A prototype instrument. *Archives of Neurology*, *52*(10), 949–954.

Martin, J. F. (2009). Article 7: Persons without the capacity to consent. In H. A. M. J. ten Have & M. S. Jean (Eds.), *The UNESCO Universal Declaration on Bioethics and Human Rights: Background, Principles and Application*. Paris: UNESCO Publishing.

Meisel, A., Roth, L. H., & Lidz, C. W. (1997). Toward a model of the legal doctrine of informed consent. *American Journal of Psychiatry*, *134*(3), 285–289. doi:10.1176/ajp.134.3.285.

Okonkwo, O., Griffith, H. R., Belue, K., Lanza, S., Zamrini, E. Y., Harrell, L. E., ... Marson, D. C. (2007). Medical decision-making capacity in patients with mild cognitive impairment. *Neurology*, *69*(15), 1528–1535. doi:10.1212/01.wnl.0000277639.90611.d9.

Orgogozo, J. M., Gilman, S., Dartigues, J. F., Laurent, B., Puel, M., Kirby, L. C., ... Hock, C. (2003). Subacute meningoencephalitis in a subset of patients with AD after Abeta42 immunization. *Neurology*, *61*(1), 46–54.

Pierce, R. (2010). A changing landscape for advance directives in dementia research. *Social Science & Medicine*, *70*(4), 623–630. doi:10.1016/j.socscimed.2009.10.037.

Porteri, C. (2018). Advance directives as a tool to respect patients' values and preferences: discussion on the case of Alzheimer's disease. *BMC Medical Ethics*, *19*, 9. doi:10.1186/s12910-018-0249-6.

Prince, M., Bryce, R., Albanese, E., Wimo, A., Ribeiro, W., & Ferri, C. P. (2013). The global prevalence of dementia: a systematic review and metaanalysis. *Alzheimer's & dementia: the journal of the Alzheimer's Association*, *9*(1), 63–75. doi:10.1016/j.jalz.2012.11.007.

Reyna, Y. Z., Bennett, M. I., & Bruera, E. (2009). Ethical and practical issues in designing and conducting clinical trials in palliative care. In: J. Addington-Hall, E. Bruera, I. J. Higginson, & S. Payne (Eds.). *Research Methods in Palliative Care*. Oxford, Oxford University Press.

Roberts, L. W. (2002). Informed consent and the capacity for voluntarism. *American Journal of Psychiatry*, *159*(5), 705–712. doi:10.1176/appi.ajp.159.5.705.

Rubinstein, E., Duggan, C., Landingham, B. V., Thompson, D., & Warburton, W. (2015). A call to action the global response to dementia through policy innovation. Report of the WISH Dementia Forum 2015. London, World Innovation Summit for Health. Retrieved from http://www.mhinnovation.net/sites/default/files/downloads/resource/WISH_Dementia_Forum_Report_08.01.15_WEB.pdf. Accessed on April 28, 2018.

Saks, E. R., Dunn, L. B., Wimer, J., Gonzales, M., & Kim, S. Y. H. (2008). Proxy consent to research: The legal landscape. *Yale Journal of Health Policy Law Ethics*, *8*(1), 37–78.

Schmidhuber, M., Haeupler, S., Marinova-Schmidt, V., Frewer, A., & Kolominsky-Rabas, P. L. (2017). Advance directives as support of autonomy for persons with dementia? A pilot study among persons with dementia and their informal caregivers. *Dementia and Geriatric Cognitive Disorders EXTRA*, *7*(3), 328–338. doi:10.1159/000479426

Scholten, M., & Gather, J. (2018). Adverse consequences of article 12 of the UN Convention on the Rights of Persons with Disabilities for persons with mental disabilities and an alternative way forward. *Journal of Medical Ethics*, *44*(4), 226–233. doi:10.1136/medethics-2017-104414.

Sessums, L. L., Zembrzuska, H., & Jackson, J. L. (2011). Does this patient have medical decision-making capacity? *Journal of the American Medical Association*, *306*(4), 420–427. doi:10.1001/jama.2011.1023.

Shepherd, V. (2016). Research involving adults lacking capacity to consent: the impact of research regulation on 'evidence biased' medicine. *BMC Medical Ethics*, *17*(1), 55. doi:10.1186/s12910-016-0138-9.

Silverman, H., Hull, S. C., & Sugarman, J. (2001). Variability among institutional review boards' decisions within the context of a multicenter trial. *Critical Care Medicine*, *29*(2), 235–241.

Silverman, H., Luce, J., & Schwartz, J. (2004). Protecting subjects with decisional impairment in research: The need for a multifaceted approach. *American Journal of Respiratory Critical Care Medicine*, *169*(1), 10–14. doi:10.1164/rccm.200303-430CP.

Slaughter, S., Cole, D., Jennings, E., & Reimer, M. A. (2007). Consent and assent to participate in research from people with dementia. *Nursing Ethics*, *14*(1), 27–40. doi:10.1177/0969733007071355.

Strech, D., Mertz, M., Knüppel, H., Neitzke, G., & Schmidhuber, M. (2013). The full spectrum of ethical issues in dementia care: systematic qualitative review. *The British Journal of Psychiatry*, *202*(6), 400–406. doi:10.1192/bjp.bp.112.116335.

Sugarman, J., McCrory, D. C., & Hubal, R. C. (1998). Getting meaningful informed consent from older adults: a structured literature review of empirical research. *Journal of the American Geriatrics Society*, *46*(4), 517–524.

Tolson, D., Fleming, A., Hanson, E., de Abreu, W., Crespo, M. L., Macrae, R., ... Routasalo, P. (2016). Achieving Prudent Dementia Care (Palliare): An International Policy and Practice Imperative. *International Journal of Integrated Care*, *16*(4), 18. doi:10.5334/ijic.2497.

Tolson, D., Rolland, Y., Andrieu, S., Aquino, J. P., Beard, J., Benetos, A., ... The International Association of Gerontology and Geriatrics/World Health Organization/Society Française de Gérontologie et de Gériatrie Task Force. (2011). International Association of Gerontology and Geriatrics: A global agenda for clinical research and quality of care in nursing homes. *Journal of the American Medical Directors Association*, *12*(3), 184–189. doi:10.1016/j.jamda.2010.12.013.

Trachsel, M., Hermann, H., & Biller-Andorno, N. (2015). Cognitive fluctuations as a challenge for the assessment of decision-making capacity in patients with dementia. *American Journal of Alzheimer's Disease & Other Dementias*, *30*(4), 360–363. doi:10.1177/1533317514539377.

Welch, M. J., Lally, R., Miller, J. E., Pittman, S., Brodsky, L., Caplan, A. L., ... Wilfond, B. (2015). The ethics and regulatory landscape of including vulnerable populations in pragmatic clinical trials. *Clinical Trials*, *12*(5), 503–510. doi:10.1177/1740774515597701.

Wendler, D. (2017). A pragmatic analysis of vulnerability in clinical research. *Bioethics*, *31*(7), 515–525. doi:10.1111/bioe.12367.

Wendler, D., & Prasad, K. (2001). Core safeguards for clinical research with adults who are unable to consent. *Annals of Internal Medicine*, *135*(7), 514–523.

West, E., Stuckelberger, A., Pautex, S., Staaks, J., & Gysels, M. (2017). Operationalising ethical challenges in dementia research-a systematic review of current evidence. *Age and Ageing*, *46*(4), 678–687. doi:10.1093/ageing/afw250.

Woods, B., & Pratt, R. (2005). Awareness in dementia: Ethical and legal issues in relation to people with dementia. *Aging & Mental Health*, *9*(5), 423–429. doi:10.1080/13607860500143125.

World Health Organization. (2015). First WHO ministerial conference on global action against dementia: meeting report. Geneva: World Health Organization. Retrieved from http://apps.who.int/iris/bitstream/handle/10665/179537/9789241509114_eng.pdf;jsessionid=4FE21E1027C6CD095390F3F7A5B21D23?sequence=1

Wortmann, M. (2012). Dementia: a global health priority - highlights from an ADI and World Health Organization report. *Alzheimer's Research & Therapy*, 21, *4*(5), 40. doi:10.1186/alzrt143.

DIET THERAPY EFFECTIVE TREATMENT BUT ALSO ETHICAL AND MORAL RESPONSIBILITY

Jasenka Gajdoš Kljusurić

ABSTRACT

Diet therapy or nutritional therapy has become a real challenge in the fight against the increasing number of modern illnesses such as obesity, diabetes, cardiovascular diseases and cancers. The scientific community has recognized the importance of studies that will support or rebut the association of certain nutrition/energy inputs with the prevention and/or improvement of certain diseases. Patient counseling is offered by medical doctors, nutritionists and dieticians, but patients often seek additional sources of information from popular media that may not be adequately scientifically supported. Whose responsibility is it when the Diet Therapy is not an effective treatment and where does the consequent ethical and moral responsibility lie?

This chapter argues for the importance of a nutritionally educated scientist evaluating the diets that are seen to be related with the health improvement also excluding diets that are mostly related to the patients' well-being as the Mediterranean, DASH (Dietary Approaches to Stop Hypertension), Ketogenic and Vegetarian diet. Diet guidelines are often explained with linguistic variables (as "reduce the input of" etc.) which can be differently perceived by the end user. The interpretation if a linguistic variable is presented using the body mass index categories using a bell-shaped curve. The preferable area fits to the linguistic variable "acceptable BMI." But also are indicated those areas which are less preferable. Those examples of information interpretations show the necessity of knowledge transfer. The quantity of information presented in diet guidelines can be experienced as a great muddle for patients; leaving them not knowing where and how to

Ethics and Integrity in Health and Life Sciences Research
Advances in Research Ethics and Integrity, Volume 4, 169–184

ISSN: 2398-6018/doi:10.1108/S2398-601820180000004011

start. So, remains the ethical and moral responsibility of all links in the chain of nutritional and diet research and recommendations. Only objective and open-minded recommendations based on the latest scientific facts can gain confidence of the social, economical, and political subjects which must put the well-being of the population uppermost in their mind.

Keywords: Diet therapy; fuzziness; linguistic variables; responsible counseling; scientific integrity; ethics

INTRODUCTION

Drug therapy is seen to be crucial in treating disease but the effectiveness of diet therapy has increased (Alvarez-Alvarez et al., 2018) and has proved to be an extremely important factor in patient treatment (Ball et al., 2018). Nutritionists and dietitians recognize that the initiator of diet therapy was Hippocrates with his statement to: "Let food be your medicine" which indicated the role of appropriate diet in a patient's healing. The word "diet" does not exclusively refer to nourishment that has the purpose of losing body weight which is how it is popularly conceived. One's diet primarily represents the regular intake of energy and diverse nutrients to maintain the life and health. Diet is a form of eating that follows certain recommendations in certain food(s) intake as well as the distribution of meals and/or snacks during the day. Some changes in our nourishment habits (diet(s)) are needed with the aim of treating a disease – it is then that we may refer to diet as therapy (Brunt, 2006; Ogden, Wood, Payne, Fouracre, & Lammyman, 2018). Medical doctors, nutritionists, and dietitians can help plan meals for the managing of symptoms of health problems as well as for maintaining good health. However, emphasized value of "information" becomes essential requirement for effective and efficient management equating knowledge with power (Davies, 1994). Presentation of any information has to be objective without personal attitude toward the presented topic. This is the ethical and moral obligation of knowledge holders. The same pattern is mapped to therapy counseling, especially to diet therapy counseling.

DIET THERAPY

During January 2018 the search engine of Web of Science[1] was searched with the keywords: diet (total: 127,186), vs. illness (1,025 papers) and health (26,537 papers). Over 120 thousand papers on these topics were published in the Journals cited in Current Contents database. Twenty percent of all scientific papers were devoted to the relation of diets with human health and, in the last five years the number has risen showing the growing interest of scientists in this topic (Fig. 1).

Food intake related to human health is seen in 25 times more papers than in relation to illness ("diet and health" vs. "diet and illness"), confirming that diet is primarily related to healing and maintaining health. Diet therapy is defined as

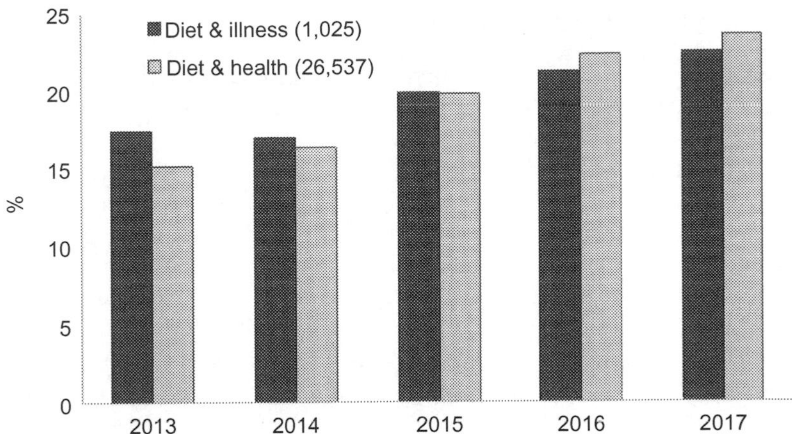

Fig. 1. Percentage of Published Papers (Same Topic) in the Last Five Years, Since 2013 to 2017. The Topic "Diet" was Refined Using the Combination Keywords "Illness" and "Health."

the dietetic treatment of various diseases and its scientific basis is food science in general, nutrition and dietetics. Diet therapy is closely coordinated with the general plan of treatment for an illness. Sometimes it is used as the principal method of treatment (in the case of diabetes see Ho, Yu, & Lee, 2017). At other times, it is the essential therapeutic background against which all other treatment (including any specific treatment therapy) must be used. Different diseases will require different directions about eating habits.

The effect of a diet therapy is based on a certain (1) energy content, (2) quantity, and (3) nutritive composition of the food at the same time as taking into account the pathogenic nature of the disease and the patient's preferences which are related to their personal tastes and ethnic customs. A diet guideline will also recommend the number of meals in the day. The majority of diets recommend at least 3 daily meals (breakfast, lunch, and dinner) and some diets add one or two snacks. A snack is distinguished form the meal based on their energy content ranged from 50 to 430 kcal (Bellisle, 2014). Any diet should aim to improve or maintain the health status of every single person.

Each year, in January the *U.S. News & World Report*[2] releases a ranking of popular diets, assessed by a panel of health experts based on seven categories. This panel consists of nationally recognized experts in food science, dietetics, nutrition, obesity, food psychology, diabetes and heart disease. They rate each diet based on following categories: (1) ease in following the recommendations; (2) ability to produce short-term weight loss; (3) ability to produce long-term weight loss; (4) the diet's nutritional completeness; (5) diet safety; (6) the diet's potential for preventing and managing diabetes; and (7) the potential for preventing and managing heart diseases (Egger, Binns, & Rossner, 2009; Eilat-Adar, Sinai, Yosefy, & Henkin, 2013; NIH, 2005). The panelists' experience is important because, as mentioned, weight loss is a secondary gain, but has to be evaluated. Knowing that in the case of the overweight or obese, a loss of just 5% of their body weight can dramatically reduce the risk of chronic illnesses such as diabetes and heart disease (Eilat-Adar et al., 2013; Juraschek, Miller, Weaver, & Appel, 2017). Safety is one of the most important characteristics because each diet should offer nutritional completeness providing sufficient calories while not falling seriously short on important nutrients or entire food groups (Braha, Cupák, Pokrivčák, Qineti, & Rizov, 2017). A safe diet will not result in malnourishment, rapid weight loss, contraindications for certain populations or existing conditions (Raphaely & Marinova, 2014). The panellists' judgment include also user's preferences about each diet's taste, ability to keep dieters from feeling hungry, the ease of its adjustment to the dieter's requirements and so on (Brunt, 2006). Exercise receives serious attention in some diets and is merely paid lip service in others, but the primary focus of a diet, after all, is food intake. Whether to exercise (how and how much) is a lifestyle decision beyond the scope of diet alone. Topping 2018's list of "best diets overall" is a tie between the Mediterranean and DASH (Dietary Approaches to Stop Hypertension) Diets: "No single diet works for everybody, but the DASH and the Mediterranean diets have the strongest biological underpinning" (Cohen, Gorski, Gruber, Kurdziel, & Rimm, 2016; Gallieni & Cupisti, 2016).

The panelists ranked the first ten diets: DASH Diet; Mediterranean diet; the Flexitarian Diet; Weight Watchers Diet; MIND Diet; TLC Diet; Mayo Clinic Diet; Ornish Diet; the Fertility Diet; and the Vegetarian Diet. Below you will find the main guidelines and features of each of the above-mentioned diets.

The DASH Diet's aim is to prevent and lower high blood pressure (hypertension). This can be achieved by eating food consisting of nutrients like potassium, calcium, protein, and fiber that are crucial to fending off or fighting high blood pressure (Gallieni & Cupisti, 2016; Park et al., 2017). The first step is the reduction of salting food (cutting back on salt). Emphasized foods to eat are fruits, veggies, whole grains, lean protein, and low-fat dairy; limiting foods that are high in saturated fat (fatty meats, full-fat dairy foods, tropical oils, and sugar-sweetened beverages and sweets). This diet presents a healthy eating pattern aimed to lower high blood pressure which is often also recommended for pregnant women (Đunđek, Gajdoš Kljusurić, Magdić, Lukinac Čačić, & Kurtanjek, 2011).

The Mediterranean Diet's aims may include weight loss but its main aim is the prevention of heart and brain disease, cancer, and diabetes prevention and control. Numerous studies have confirmed that residents in countries bordering the Mediterranean Sea live longer and suffer less than most Americans from cancer and cardiovascular ailments (Juraschek et al., 2017; Park et al., 2017). Features of the Mediterranean diet besides the specific lifestyle is a low consumption of red meat, sugar, saturated fats and a high consumption of vegetable, nuts and other healthful foods. This diet is dominated by Mediterranean foodstuffs, such as olive oil, with fish as the preferred animal protein, vegetables and fruits, dairy, and a moderate intake of wine with meals.

Alongside this diet daily physical activity is recommended (Alvarez-Alvarez et al., 2018).

The Flexitarian Diet presents a flexible vegetarian diet without entirely giving up meat, assuring weight loss and optimal health. This diet is flexible even considering the meat consumption "...you don't have to eliminate meat completely to reap the health benefits associated with vegetarianism – you can be a vegetarian most of the time, but still chow down on a burger or steak when the urge hits" (Blatner, 2010). Studies show that flexitarians weigh on average 15% less, have a lower rate of heart disease, diabetes and cancer and live nearly four years longer than their carnivorous counterparts (Raphaely & Marinova, 2014).

The MIND Diet, which stands for "Mediterranean-DASH Intervention for Neurodegenerative Delay," was developed by Martha Clare Morris, a nutritional epidemiologist at Rush University Medical Center (Morris, Tangney, Wang, Sacks, Barnes et al., 2015; Morris, Tangney, Wang, Sacks, Bennett, et al., 2015). It may lower the risk of mental decline with a new hybrid of two balanced, heart-healthy diets – even without rigidly sticking to it, DASH and the Mediterranean diet – and zeroes in on the foods in each that specifically affect brain health (Marcason, 2015). Some research has shown the relation of the MIND diet with prevention of Alzheimer's disease with brain-healthy foods (Egger et al., 2009; Marcason, 2015).

The central point of this diet relies on eating from 10 brain-healthy food groups: green leafy vegetables in particular, all other vegetables, nuts, berries, beans, whole grains, fish, poultry, olive oil, and wine. This diet also highlights the five unhealthy groups and foods whose input needs to be minimized: such as red meats, butter and stick margarine, cheeses, pastries and sweets, and fried or fast food.

The Therapeutic Lifestyle Changes (TLC) Diet aims to reduce high cholesterol. This diet is endorsed by the American Heart Association as a heart-healthy regimen that can reduce the risk of cardiovascular disease (NIH, 2005). The key is cutting back sharply on fat, particularly saturated fat which bumps up bad cholesterol increasing the risk of heart attack and stroke. This diet cuts saturated fat (such as fatty meat, whole-milk dairy, and fried foods) followed by strictly limiting daily dietary cholesterol intake and increasing fiber intake, which can help people manage high cholesterol, often without medication (Eilat-Adar et al., 2013; NIH, 2005).

The Volumetrics Diet is characterized by plentiful, nutrient-dense food and is primarily aimed at weight loss. This diet allows one to eat the same volume of food, but since some foods are less energy dense than others, such foods (high nutrient dense and less energy dense) will fill the plate. Low-density foods, which are low in calories but high-volume like fruits and vegetable, help one to feel full and satisfied while dropping the pounds (Vogliano et al., 2015).

The Mayo Clinic Diet results in weight loss and a healthier lifestyle. The idea is to recalibrate present eating habits, breaking bad ones and replacing them with good ones with the help of the Mayo Clinic's unique food pyramid (Brunt, 2006). This food guide pyramid emphasizes fruits, vegetables, and whole grains. In general, these foods also have low-energy density.

The Ornish Diet can be tailored to lose weight, prevent or control diabetes and heart disease, lower blood pressure and cholesterol, prevent and treat prostate or breast cancer, and reverse cellular aging by lengthening telomeres (Nosova, Conte, & Grenon, 2015).

The Fertility Diet is claimed to boost ovulation and improve fertility. Chavarro, Willett, and Skerrett (2007) published: "The Fertility Diet: Groundbreaking Research Reveals Natural Ways to Boost Ovulation and Improve Your Chances of Getting Pregnant" showing the importance of the diet in aspects related to fertility. The results of the study showed that women who consumed "good" fats, whole grains, and plant protein improved their egg supply, while those who ate "bad" fats, refined carbohydrates and red meat increased the risk for ovulatory infertility. This research also suggests that full-fat dairy products are good for fertility compared with skim milk and sugary sodas.

The Vegetarian Diet can include weight loss but is proven to assure heart health and diabetes prevention or control (Garbett, Garbett, & Wendorf, 2016). This diet shows that it is possible to cook up a perfectly healthy, meat-free menu that supports weight loss and reduces the risk of heart disease and diabetes (Garbett et al., 2016; Ho et al., 2017; Rosenfeld & Burrow, 2017).

A large population-based Australian cohort study included the sector of the population aged 45 and upward and investigated the vegetarian diet and all-cause mortality (Mihrshahi et al., 2017). Their findings were based on the hypothesis that a vegetarian diet is thought to have health benefits including reductions in type 2 diabetes (Kahleova & Pelikanova, 2017), hypertension and obesity (Aggarwal et al., 2015). However, they point out the doubt in the suggestions that vegetarians tend to have lower mortality rates when compared with non-vegetarians, highlighting an important factor that in most studies was not taken into account – they are not population-based and other healthy lifestyle factors may have confounded protective effects. Their study included a cohort study of 267,180 men and women aged ≥45 years in New South Wales, Australia. The final results showed that there was "We found no evidence that following a vegetarian diet, semi-vegetarian diet or a pesco-vegetarian diet has an independent protective effect on all-cause mortality" (Mihrshahi et al., 2017).

Each diet has to follow some specific recommendations, and to help their users in menu planning, different tools are used. For instance, one diet that aims to help with weight loss and living healthier is the Weight Watchers Diet. This diet assigns every food and beverage a SmartPoints value, based on its nutritive composition, and the idea is to use food and meals with lower point values which includes a lower intake of saturated fats and sugars as well of a higher intake of proteins and foods of high nutritional density (higher amounts of saturated fat and sugar increase the point value; higher amounts of protein and nutritional density bring the point value down).

From just a few of these mentioned diets, there is something in common, and that is the beneficiaries' well-being, which is manifested by an adequate physical and health status. Some studies confirm the more we change our diet, the more health benefits can be reaped – at any age (Cohen et al., 2016). So, the diet can be used for only looking to lose a few pounds as well as to reverse some disease – which research shows may be possible at the rigorous end of this diet's spectrum of choices.

DIET THERAPY AND SCIENCE

Some diets, as Ketogenic diet, will not appear on a list of mostly efficient and used diets. The Ketogenic diet is based on the biochemistry in the human body where ketone bodies play an important role beside glucose as oxidizable substrates and an energy source for the brain (Morris, 2005; Roberts et al., 2017); in adult humans, high ketone body concentrations are found during fasting and on a high-fat diet. Lange and his coworkers (2017) pointed out that the content of ketone bodies plays an important role – if they are present in sufficient concentrations that saturate the metabolism, then they can support the basal neuronal energy and nearly a half of the oxidative activity-dependent neuronal needs (Lange, Lange, Makulska-Gertruda, Nakamura, & Hauser, 2017).

In normal circumstances, the body as the primary source of energy uses carbohydrates. Nowadays, in human nutrition, carbohydrates present more than 3/5 in the daily energy intake what encourages signaling of insulin leading to lipid metabolism suppression, so ketogenesis may be more evolutionarily discordant than high-fat diets (Lange et al., 2017). The ketogenic diet is very high in fat and low in carbohydrates and is believed to simulate the effects of starvation by primarily metabolizing fat as energy supply. If we exclude carbohydrates from the diet, the body is forced to find an alternative source of energy. One of these energy sources are free fatty acids. The brain cannot use free fatty acids as propellant fuel; however, the ketone bodies can be used. Ketone bodies are by-product of incomplete degradation of fatty acids in the liver. When they are rapidly produced, they accumulate in the blood and a condition called ketosis occurs. Since the body now has again a source of energy (adenosine triphosphate, ATP), there is no longer a need for carbohydrates, so gluconeogenesis (amino acid glucose production) is stopped. Consequently, the proteins remain spared of degradation (otherwise it would be used for energy purposes) – which is the main principle of the ketogenic diet (Ingram et al., 2006). Those facts are the reason why the ketogenic diet has now become an established and effective nonpharmacological treatment for epilepsy (Kumada, Imai, Takahashi, Nabatame, & Oguni, 2018; Williams & Mackenzie, 2017), Alzheimer's disease (Lange et al., 2017), anorexia nervosa (Scolnick, 2017), inherited metabolic disease (Morris, 2005), dementia (Taylor, Sullivan, Mahnken, Burns, & Swerdlow, 2018), and potential longevity and healthiness (Roberts et al., 2017). The human brain requires approximately 20% of the entire body's supply of energy. In this diet are several mechanisms underlying the anticonvulsive effects of ketone bodies, including changes in ATP production, altered brain pH affecting neuronal excitability, direct inhibitory effects of ketone bodies or fatty acids on ion channels, and shifts in amino acid metabolism (Lange et al., 2017). The ketogenic diet is likely to have different mechanisms in different diseases (Bough & Rho, 2007).

To get an overview of the different scientific interest, the search engine WoS was used to show different interest (Fig. 2) and interest decrease and increase (Fig. 3) for Mediterranean, DASH, Ketogenic, and Vegetarian diets.

The interest about the well-being of Ketogenic and Vegetarian diets has risen in the last five years, while the DASH diet has appeared in some criticizing studies (Bloch, 2017) presenting less enthusiasm and analyzing studies (Steinberg, Bennett, & Svetkey, 2017) which have shown that the DASH diet is underutilized among the tens of millions of US citizens who have hypertension and prehypertension. Bloch (2017) has highlighted a number of potential barriers to the more widespread adoption of the DASH eating plan as lack of knowledge and training in medical nutrition of the public. The main approach in the DASH diet treatment of hypertension and chronic cardiovascular insufficiency comprises foods containing no more than 2–3 g of common salt, enriched in potassium and magnesium salts and vitamins, and containing physiologically normal amounts of proteins, fats, and carbohydrates. In addition, a magnesium diet is prescribed briefly from time to time for the benefit of the depressor action of its magnesium salts (Tangvoraphonkchai & Davenport, 2018). The DASH diet guidelines emphasize increased consumption of fruits (preferring fruits as banana as good source of potassium), vegetables, and low-fat dairy products. Recommended are also whole grains, poultry, fish and nuts with emphasis on reduction in salt, fats, red meat, sweets, and sugar-containing beverages. If the population is not educated, how can they decide how many nuts is preferable or what is considered under reduction of fats? Does it mean not to use them, or are included salad dressings as well?

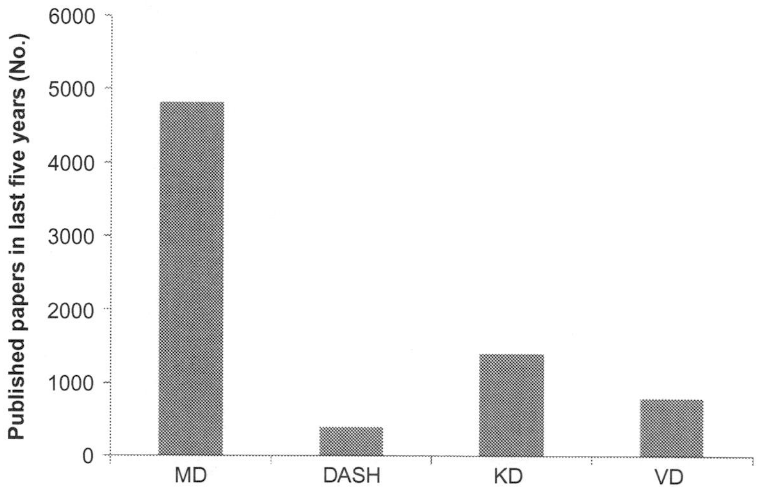

Fig. 2. Published Papers in the Last Five Years (Since 2013 to 2017) with the Topic of the Most-mentioned Diets In Relation To Health as Mediterranean Diet (MD), DASH Diet (DASH), Ketogenic Diet (KD), and Vegetarian Diet (VD).

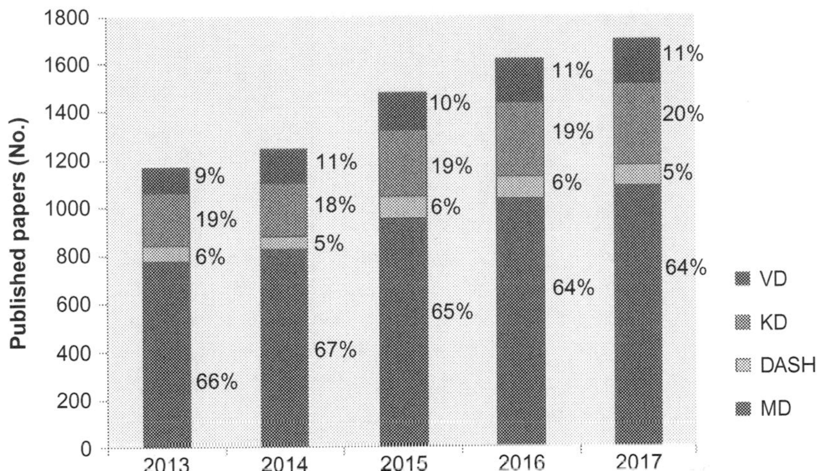

Fig. 3. Share of the Most-mentioned Diets In Relation To Health in the Last Five Years for Mediterranean diet (MD), DASH Diet (DASH), Ketogenic Diet (KD), and Vegetarian Diet (VD).

If we consider the diabetic diet, their intake of the readily soluble carbohydrates is restricted (sugar, glucose), the diet should contain 70–80 g of fat (including 30 g vegetable oil), 100–120 g of proteins (chiefly those possessing lipotropic properties), and vitamins A, B, and C (Đunđek et al., 2011).

The Vegetarian diet seems to be the simplest including plant-based foods, but the food choices in a vegetarian diet require special care in meal planning because their dietary habits range from semi-vegetarians, lacto-ovo vegetarians to strict vegans. However, an adequate vegetarian diet can be appropriate for all the stages of the life cycle, including pregnancy, lactation, infancy, childhood, adolescence, as well as for athletes (ADA, 2009). The number of vegetarians and vegans in society is growing due to environmental, social, and health concerns. The initiative "Green Monday" began in 2003, promoting a diet to avoid meat consumption for at least one day per week. This promotion has been worldwide (ZeGeVege, 2018). The basic intention of a vegetarian diet is to improve personal health and to reduce the ecological footprint of the usual food choices (Kemp, Insch, Holdsworth, & Knight, 2010; Rosenfeld & Burrow, 2017). The study of Marlow et al. (2009) argued that the omnivore diet requires 2.9 times more water, 2.5 times more primary energy, 13 times more fertilizer, and 1.4 times more pesticides than the vegetarian diet (Marlow et al., 2009). But additional attention is needed because the list of critical nutrients (missing?) in a vegetarian diet is composed of proteins, n-3 fatty acids, two very important vitamins, vitamin D and B_{12}, and minerals as calcium, zinc, iodine and iron (Orešković, Gajdoš Kljusurić, & Šatalić, 2015).

SIMPLICITY OR COMPLEXITY OF THE USE OF DIET THERAPY

The certain element of diet therapy is the factor of changing some eating habits and lifestyle using scientifically proven methods to make us to "feel better, live longer, lose weight and gain health" – which is one of the main ideas of the Ornish diet (Nosova et al., 2015). All these mentioned goals are linguistically variable (better, longer etc.) whose values are words or sentences in a natural or artificial language. For example, "good health status" is a linguistic variable if its values are linguistic rather than numerical, that is, healthy, not healthy, very healthy, quite healthy, unwell, not very diseased and not very well, etc., rather than on the scale from 1 to 10 – give a numeric value of the well-being. Pathological and clinical measures can be used to measure health status and such measures include determining (1) signs as body temperature, blood pressure, etc., (2) symptoms using diseases-specific checklists and some disease-related indexes (as Charlson Index, ICED index). Observing the linguistic variable "healthy," it would be expected to have normal "signs" (temperature, blood pressure, etc.) with the absence of all other indicators. Now, these important points require some clarification. Is a patient "unhealthy" having all signs in the normal range, but having a small kidney stone that has not caused any problems? This is an example of a linguistic variable; providing those information (all normal signs and a small kidney stone) will for one parson be experienced as "healthy," while another will perceive it as "quite healthy." Thus measurement of "health status" has long been disputed and the numerical values are often assigned subjectively – which keeps them close to "linguistic variable." Even BMI as a measure has its critics but some ranges are necessary to give an insight in the problematic (Gajdoš Kljusurić, Rumora, & Kurtanjek, 2012; Singh & Dubey, 2017). Such linguistic variable for the Body Mass Index, "normal BMI" and "good health status" could be presented as following curves (Fig. 4):

If the calculated BMI for a parson is 25 kg m^{-2}, is this person Overweight or Obese? The crisp set (dashed line) will group this person in the set Overweight & Obese, while the membership in a fuzzy set is quite flexible (Fig. 4). In the case of the BMI equal to 25, the person will still be a member of the group "normal BMI," with a partial (very small) membership to the set Overweight and Obese. This is related to the "membership" that is in a crisp set defined as a membership (μ=1) or no membership to a set (μ=0), while the fuzzy set allows a membership in the range from over zero to 1 ($0<\mu\leq 1$). Such "fuzziness" is often used in diet therapies. Advice such as consume "more" vegetable, "less" fat, etc., are confusing for the majority of the public, and as presented with the "fuzzy" theory, the perception of linguistic variables can be perceived differently. The advice of "increase" physical activity can be perceived from activities: "walk at least 1000 steps per day" to "go to the gym 5 times per week."

The goal of the public health sector, being a medical doctor, nutritionist or dietitian is to achieve the green section for the BMI and well-being of the population (Fig. 4). Any deviations are related to the increase of modern illnesses

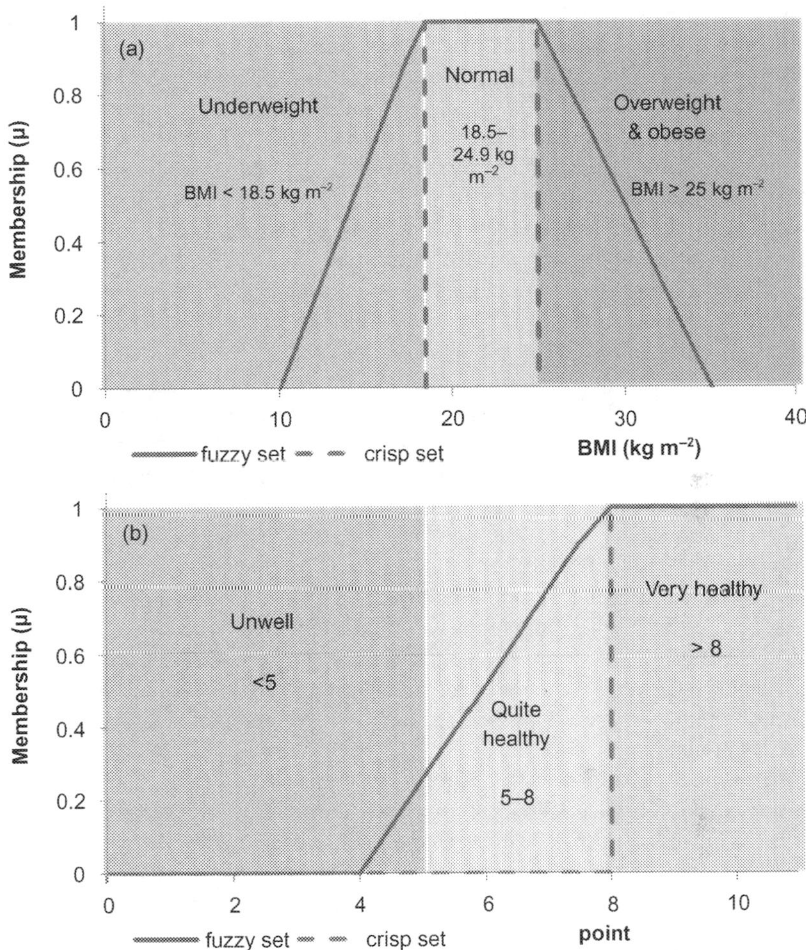

Fig. 4. Membership to the Linguistic Variable[3] (1) "Normal BMI" and (2) "Good Health Status."

which are highly correlated to very high/low BMI and bad nutritive habits (NIH, 2005).

It is obvious that such advice can be confusing for the patient. Taking into account that the diet should be individualized, diet therapy counseling becomes complicated and time consuming. All of us have a subjective attitude toward food, diet, and lifestyle in general. But diet therapy must remain objective with minor changes in relation to the basic diet guidelines because the stake is great – the human health.

Eating behavior is also a psychological process while food intake is influenced by cognition and the food environment (de Beaufort, 2014). The problem could be the media framing information as personal, "lifestyle" issues emphasizing individual ingredients or components more than overall diet. Illnesses as obesity, diabetes type 2, coronary heart disease, or even preventing bowel cancer is set predominantly in the "lifestyle" frame rather than looking at the broader initiators of nutritional changes (as social, economic, or political). Medical doctors, nutritionists, and dietitians' have the responsibility to be the first router in enabling people to choose health by improving their eating habits, criticizing scientifically unfounded weight loss diets, and promoting diets which are scientifically supported to improve a certain medical illness where the secondary effect may be the loss of body mass (Wells, 2017).

Kim and Ham (2016) investigated the nutritional information as a corporate social responsibility initiative that shows that the way of presentation of nutritional information influence on Consumers loyalty. The patient trust related to diet advices is related with different factors where one of them can even be the body mass index of their primary care adviser (Bleich, Gudzune, Bennett, Jarlenski, & Cooper, 2013). Nowadays we witness the rise of diet guidelines which contain the list of "good" and "bad foods," what is highlighted as one of "top ten red flags" for misleading claims (Bellow & Moore, 2013). Considering the Mediterranean diet, on the list of "good food" appears "glass of wine." Without additional information obtained from food science/nutrition educated person, wine consumption could be abused. Responsible advisors will clarify the amount of "glass of wine" explaining that consumption of red wine is preferable and the amount depends on the gender; 0.1 L for women and 0.2 L for men, with an emphasis on the fact that the consumption of wine is recommended with the most plenteous meal(s), which is lunch or dinner (Alvarez-Alvarez et al., 2018; Urquiaga et al., 2010).

The fuzziness of information can mislead even the experts. Living in the age of easy accessibility of information, even the diet therapy advisors must be careful in selection of nutrition information. A person on a diet without expert supervision will reach for nutrition information from different sources (advertisements, websites, television, radio, newspapers) increasing the likelihood of exposure for nutrition misinformation not supported by science, misleading and incomplete. Consultation of a person on a diet by an educated staff will provide accurate nutrition information that are science-based and peer reviewed.

The classic version of Hippocratic Oath[4] pointed following: "I will apply dietetic measures for the benefit of the sick according to my ability and judgment; I will keep them from harm and injustice." In the *Modern Version of the Hippocratic Oath[4]* can be found: "I will apply, for the benefit of the sick, all measures which are required, avoiding those twin traps of overtreatment and therapeutic nihilism."

Following the classic or the modern version of Hippocratic Oath, we cannot escape responsibility, which implies objectively consideration of a broad spectrum of scientific research that accompanies nutritional therapy as well. Patients and their well-being must be the main focus, regardless of our attitudes, beliefs, and/or preferences. Only with such working attitude in relation to the diseased, a medical doctor, nutritionist, or dietician can "avoid those twin traps of overtreatment and therapeutic nihilism" from the Hippocratic Oath.

Is it easier to print a diet guideline in a form of a list of desirable foods and give it to a person on diet – yes. Is such approach less time consuming – yes. Does it also save money – it depends. From the point of view of a diet advisor, in a short-term yes, but not in the long run. Every minute more spend in clarifying uncertainties in a diet, compatible foods, and similar is a small step closer to possible health improvement. Health is priceless and sometimes irretrievable, as realized by those who have lost it, as the proverb says: "A healthy person has a thousand wishes, a sick person only one."

The calling of a medical doctor, nutritionist, or dietician regarding diet therapy is a very responsible profession, morally and ethically, but as it often happens in life, sometimes it seems easier, faster, and more effective to use the shortcuts. This lies in our human nature, but regardless if a diet is intended to weight loos or as additional treatment during an illness, the effectiveness of the treatment is an ethical and moral responsibility.

We have chosen these calls in order to serve the human population to maintain or even improve their well-being, and this chapter is only a small reminder how we would not forget it.

NOTES

1. Retrieved from http://apps.webofknowledge.com. Accessed on February 1, 2018.
2. Retrieved from https://www.usnews.com/info/blogs/press-room/articles/2018-01-03/us-news-reveals-best-diets-rankings-for-2018. Accessed on February 1, 2018.
3. Presented membership functions are modeled for the purposes of this chapter by the author of the chapter and are nowhere else published in this form.
4. Retrieved from https://www.medicinenet.com/

REFERENCES

ADA. (2009). Position of the American Dietetic Association: Vegetarian diets. *Journal of American Dietetic Association*, *1009*(7), 1266–1282.

Aggarwal, H., Kaur, H., Saklani, R., Saba, N., Choudhary, S., Dogra, S., … Gupta, S. K. (2015). Prevalence of obesity and associated hypertension and diabetes in Delhi, metropolitan city of India. *Indian Journal of Medical Specialities*, *6*(3), 82–87. doi:10.1016/j.injms.2015.06.005.

Alvarez-Alvarez, I., Zazpe, I., Pérez de Rojas, J., Bes-Rastrollo, M., Ruiz-Canela, M., Fernandez-Montero, A., … Martínez-González, M. A. (2018). Mediterranean diet, physical activity and their combined effect on all-cause mortality: The Seguimiento Universidad de Navarra (SUN) cohort. *Preventive Medicine*, *106*(1), 45–52. doi:10.1016/j.ypmed.2017.09.021.

Ball, L. E., Sladdin, I. K., Mitchell, L. J., Barnes, K. A., Ross, L. J., & Williams, L. T. (2018). Quality of development and reporting of dietetic intervention studies in primary care: a systematic review of randomised controlled trials. *Journal of Human Nutrition and Dietetics*, *31*(1), 47–57. doi:10.1111/jhn.12526.

Bellisle, F. (2014). Meals and snacking, diet quality and energy balance. *Physiology and Behaviour*, *134*(12), 38–43. doi:10.1016/j.physbeh.2014.03.010.

Bellows, L., & Moore, R. (2013). Nutrition misinformation: How to identify fraud and misleading claims. *Food and Nutrition Series/Health*, *9*(13), 1–4.

Blatner, D. J. (2010). *The Flexitarian Diet: The mostly vegetarian way to lose weight, be healthier, prevent disease, and add years to your life*. McGraw Hill Professional. ISBN: 978-0-07-174579-6.

Bleich, S. N., Gudzune, K. A., Bennett, W. L., Jarlenski, M. P., & Cooper, L. A. (2013). How does physician BMI impact patient trust and perceived stigma? *Preventive Medicine*, *57*(2), 120–124. doi:10.1016/j.ypmed.2013.05.005.

Bloch, M. J. (2017). The dietary approaches to stop hypertension (DASH) diet—Promise unmet. *Journal of the American Society of Hypertension*, *11*(6), 323–324. doi:10.1016/j.jash.2017.04.011.

Bough, K. J., & Rho, J. (2007). Anticonvulsant mechanisms of the ketogenic diet. *Epilepsia*, *48*(1), 43–58.

Braha, K., Cupák, A., Pokrivčák, J., Qineti, A., & Rizov, M. (2017). Economic analysis of the link between diet quality and health: Evidence from Kosovo. *Economics & Human Biology*, *27*(Part A), 261–274. doi:10.1016/j.ehb.2017.08.003.

Brunt, A. R. (2006). Mayo Clinic healthy weight for everybody. Book review. *Journal of Nutrition Education and Behavior*, *38*(1), 65. doi:10.1016/j.jneb.2005.11.004.

Chavarro, J. E., Willett, W. C., & Skerrett, P. J. (2007). The fertility diet: Groundbreaking research reveals natural ways to boost ovulation & improve your chances of getting pregnant. Columbus, OH: McGraw-Hill, ISBN: 978-0071494793.

Cohen, J. F., Gorski, M. T., Gruber, S. A., Kurdziel, L. B., & Rimm, E. B. (2016). The effect of healthy dietary consumption on executive cognitive functioning in children and adolescents: a systematic review. *British Journal of Nutrition*, *116*(6), 989–1000. doi:10.1017/S0007114516002877.

Davies, S. (1994). Introduction; Information, Knowledge and Power. *IDS Bulletin*, *25*(2), 1–13. doi:10.1111/j.1759-5436.1994.mp25002001.x.

de Beaufort, I. (2014). 'Please, sir, can I have some more?' Food, lifestyle, diets: Respect and moral responsibility. *Best Practice & Research Clinical Gastroenterology*, *28*(2), 235–245. doi:10.1016/j.bpg.2014.02.001.

Đunđek, S., Gajdoš Kljusurić, J., Magdić, D., Lukinac Čačić, J., & Kurtanjek, Ž. (2011). Optimisation of the Daily Nutrient Composition of Daily Intakes During Gestation. *Croatian Journal of Food Technology, Biotechnology and Nutrition*, *6*(1–2), 45–51.

Egger, G., Binns, A., & Rossner, S. (2009). The emergence of "lifestyle medicine" as a structured approach for management of chronic disease. *Medical Journal of Australia*, *190*(3), 143–145.

Eilat-Adar, S., Sinai, T., Yosefy, C., & Henkin, Y. (2013). Nutritional recommendations for cardiovascular disease prevention. *Nutrients*, *5*(9), 3646–3683. doi:10.3390/nu5093646.

Gajdoš Kljusurić, J., Rumora, I., & Kurtanjek, Ž. (2012). Application of fuzzy logic in diet therapy – Advantages of application. In E. P. Dadios (Ed.), *Fuzzy Logic – Emerging Technologies and Applications*. Rijeka: InTech.

Gallieni, M., & Cupisti, A. (2016). DASH and Mediterranean Diets as nutritional interventions for CKD patients. *American Journal of Kidney Diseases*, *68*(6), 828–830. doi:10.1053/j.ajkd.2016.09.001.

Garbett, T. M., Garbett, D. L., & Wendorf, A. (2016). Vegetarian diet: A prescription for high blood pressure? A systematic review of the literature. *The Journal for Nurse Practitioners*, *12*(7), 452–458.e6. doi:10.1016/j.nurpra.2016.04.013.

Ho, C. P., Yu, J. H., & Lee, T. J. F. (2017). Ovo-vegetarian diet is associated with lower systemic blood pressure in Taiwanese women. *Public Health*, *153*(12), 70–77. doi:10.1016/j.puhe.2017.07.032.

Ingram, D. K., Zhu, M., Mamczarz, J., Zou, S., Lane, M. A., Roth, G. S., & de Cabo, R. (2006). Calorie restriction mimetics: an emerging research field. *Aging Cell*, *5*(2), 97–108.

Juraschek, S. P., Miller, E. R., Weaver, C. M., & Appel, L. J. (2017). Effects of sodium reduction and the DASH diet in relation to baseline blood pressure. *Journal of the American College of Cardiology*, *70*(23), 2841–2848. doi:10.1016/j.jacc.2017.10.011.

Kahleova, H., & Pelikanova, T. (2017). *Vegetarian diets in people with type 2 diabetes in vegetarian and plant-based diets in health and disease prevention*. F. Mariotti (Ed.), (pp. 369–393). Elsevier London Wall: Academic Press. doi:10.1016/B978-0-12-803968-7.00021-6.

Kemp, K., Insch, A., Holdsworth, D. K., & Knight, J. G. (2010). Food miles: Do UK consumers actually care? *Food Policy*, *35*(6), 504–513. doi:10.1016/j.foodpol.2010.05.011.

Kim, E., & Ham, S. (2016). Restaurants' disclosure of nutritional information as a corporate social responsibility initiative: Customers' attitudinal and behavioral responses. *International Journal of Hospitality Management*, *55*(C), 96–106. doi:10.1016/j.ijhm.2016.02.002.

Kumada, T., Imai, K., Takahashi, Y., Nabatame, S., & Oguni, H. (2018). Ketogenic diet using a Japanese ketogenic milk for patients with epilepsy: A multi-institutional study. *Brain and Development*, *40*(3), 188–195. doi:10.1016/j.braindev.2017.11.003.

Lange, K. W., Lange, K. M., Makulska-Gertruda, E., Nakamura, Y., & Hauser, J. (2017). Ketogenic diets and Alzheimer's disease. *Food Science and Human Wellness*, *6*(1), 1–9. doi:10.1016/j.fshw.2016.10.003.

Marcason, W. (2015). What are the components to the MIND diet? *Journal of the Academy of Nutrition and Dietetics*, *115*(10), 1744. doi:10.1016/j.jand.2015.08.002.

Marlow, H. J., Hayes, W. K., Soret, S., Carter, R. L., Schwab, E. R., & Sabaté, J. (2009). Diet and the environment: Does what you eat matter? *American Journal of Clinical Nutrition*, *89*(5), 1699S–1703S. doi:10.3945/ajcn.2009.26736Z.

Mihrshahi, S., Ding, D., Gale, J., Allman-Farinelli, M., Banks, E., & Bauman, A. E. (2017). Vegetarian diet and all-cause mortality: Evidence from a large population-based Australian cohort – the 45 and Up Study. *Preventive Medicine*, *97*(4), 1–7. doi:10.1016/j.ypmed.2016.12.044.

Morris, A. A. (2005). Cerebral ketone body metabolism. *Journal of Inherited Metabolic Disease*, *28*(2), 109–121.

Morris, M. C., Tangney, C. C., Wang, Y., Sacks, F. M., Barnes, L. L., Bennett, D. A., & Aggarwal, N. T. (2015). MIND diet slows cognitive decline with aging. *Alzheimer's & Dementia*, *11*(9), 1015–1022. doi:10.1016/j.jalz.2015.04.011.

Morris, M. C., Tangney, C. C., Wang, Y., Sacks, F. M., Bennett, D. A., & Aggarwal, N. T. (2015). MIND diet associated with reduced incidence of Alzheimer's disease. *Alzheimer's & Dementia*, *11*(9), 1007–1014. doi:10.1016/j.jalz.2014.11.009.

NIH, National Heart, Lung, and Blood Institute. (2005). Your Guide to Lowering Your Cholesterol With TLC. U.S. Department of Health and Human Services, Retrieved from http://www.nhlbi.nih.gov/health/public/heart/chol/chol_tlc.pdf

Nosova, E. V., Conte, M. S., & Grenon, S. M. (2015). Advancing beyond the "heart-healthy diet" for peripheral arterial disease. *Journal of Vascular Surgery*, *61*(1), 265–274. doi:10.1016/j.jvs.2014.10.022.

Ogden, J., Wood, C., Payne, E., Fouracre, H., & Lammyman, F. (2018). 'Snack' versus 'meal': The impact of label and place on food intake. *Appetite*, *120*, 666–672. doi:10.1016/j.appet.2017.10.026.

Orešković, P., Gajdoš Kljusurić, J., & Šatalić, Z. (2015). Computer-generated vegan menus: The importance of FCDB choice. *Journal of Food Composition and Analysis*, *37*, 112–118. doi:10.1016/j.jfca.2014.07.002.

Park, Y. M., Steck, S. E., Fung, T. T., Zhang, J., Hazlett, L. J., Han, K., … Merchant, A. T. (2017). Mediterranean diet, dietary approaches to stop hypertension (DASH) style diet, and metabolic health in U.S. adults. *Clinical Nutrition*, *36*(5), 1301–1309. doi:10.1016/j.clnu.2016.08.018.

Raphaely, T., & Marinova, D. (2014). Flexitarianism: Decarbonising through flexible vegetarianism. *Renewable Energy*, *67*, 90–96. doi:10.1016/j.renene.2013.11.030.

Roberts, M. N., Wallace, M. A., Tomilov, A. A., Zhou, Z., Marcotte, G. R., Tran, D., … Lopez-Dominguez, J. A. (2017). A Ketogenic Diet extends longevity and healthspan in adult mice. *Cell Metabolism*, *26*(3), 539–546.e5. doi:10.1016/j.cmet.2017.08.005.

Rosenfeld, D. L., & Burrow, A. L. (2017). Vegetarian on purpose: Understanding the motivations of plant-based dieters. *Appetite*, *116*, 456–463. doi:10.1016/j.appet.2017.05.039.

Scolnick, B. (2017). Ketogenic diet and anorexia nervosa. *Medical Hypotheses*, *109*, 150–152. doi:10.1016/j.mehy.2017.10.011.

Singh, M. P., & Dubey, S. K. (2017). Recommendation of diet to anaemia patient on the basis of nutrients using AHP and fuzzy TOPSIS approach. *International Journal of Intelligent Engineering and Systems*, *10*(4):100–108. doi:10.22266/ijies2017.0831.11.

Steinberg, D., Bennett, G. G., & Svetkey, L. (2017). The DASH diet, 20 years later. *JAMA*, *317*(15), 1529–1530. doi:10.1001/jama.2017.1628.

Tangvoraphonkchai, K., & Davenport, A. (2018). Magnesium and cardiovascular disease. *Advances in Chronic Kidney Disease*, *25*(3), 251–260. doi:10.1053/j.ackd.2018.02.010

Taylor, M. K., Sullivan, D. K., Mahnken, J. D., Burns, J. M., & Swerdlow, R. H. (2018). Feasibility and efficacy data from a ketogenic diet intervention in Alzheimer's disease. *Alzheimer's & Dementia: Translational Research & Clinical Interventions*, *4*, 28–36. doi:10.1016/j.trci.2017.11.002.

Urquiaga, I., Strobel, P., Perez, D., Martinez, C., Cuevas, A., Castillo, O., ... Leighton, F. (2010). Mediterranean diet and red wine protect against oxidative damage in young volunteers. *Atherosclerosis*, *211*(2), 694–699. doi:10.1016/j.atherosclerosis.2010.04.020.

Vogliano, C., Brown, K., Miller, A. M., Green-Burgeson, D., Copenhaver, A. A., & Schmidt, J. (2015). Plentiful, nutrient-dense food for the world: A guide for registered dietitian nutritionists. *Journal of the Academy of Nutrition and Dietetics*, *115*(12), 2014–2018. doi:10.1016/j.jand.2015.06.367.

Wells, R. (2017). Mediating the spaces of diet and health: A critical analysis of reporting on nutrition and colorectal cancer in the UK. *Geoforum*, *84*, 228–238. doi:10.1016/j.geoforum.2016.05.001.

Williams, T. J., & Mackenzie, C. C. (2017). The role for ketogenic diets in epilepsy and status epilepticus in adults. *Clinical Neurophysiology Practice*, *2*, 154–160. doi:10.1016/j.cnp.2017.06.001.

ZeGeVege. (2018). Green Monday – vegetarian menus. Retrieved from http://www.zeleni-ponedjeljak.com

THE MISMATCH OF NUTRITION AND "MEDICAL PRACTICE": THE WAYWARD SCIENCE OF NUTRITION IN HUMAN HEALTH

T. Colin Campbell and T. Nelson Campbell

ABSTRACT

Nutrition, as a science, is poorly understood, both professionally and publicly. The confusion that surrounds this science makes it very difficult, if not impossible, to formulate public health policy, which creates opportunities for political manipulation and control. Nutrition, for a century or more, has been variously described as a summation of the physiological and biochemical properties of individual nutrients in food rather than the whole food itself. This infers that isolated nutrients in supplements will function in the same way as nutrients in food. It also infers that removing or minimizing "undesirable" nutrients from food will make the food more healthful. This arises from the highly reductionist way that we focus on individual nutrients minus their natural context, both the context within the foods of which they are a part and the context within biological systems where they function. The shortcomings of this belief system may be illustrated by hugely costly mistakes made in the past, even more than a century ago, that corrupt current practices. Such mistakes have become so embedded in the contemporary narrative on nutritional science, both fundamentally and practically, that we fail to recognize the damage they continue to cause.

Alternatively, when nutritional effects are considered more within their natural contexts, that is, more wholistically, then it helps to explain, for example, the remarkable ability of nutrition, as provided by a whole food plant-based diet, to prevent even to cure varied types of cardiovascular disease. Furthermore,

Ethics and Integrity in Health and Life Sciences Research
Advances in Research Ethics and Integrity, Volume 4, 185–201

ISSN: 2398-6018/doi:10.1108/S2398-601820180000004012

the breadth of this nutritional effect for a wide variety of illnesses and diseases suggests that nutrition, properly provided by a whole food plant-based diet, is more efficacious than a combination of all the contemporary pills and procedures combined. It also suggests that genetic determinism is not the explanation for disease that is widely advanced. And finally, among still more consequences, there are many societal outcomes that can be substantially mitigated, including the escalating cost of health care and the dangerously increasing array of destructive practices that damage the environment. Many of the momentous health, economic, environmental and sociopolitical problems currently faced may be traced to a misunderstanding of the effects of food and nutrition. The task therefore is how to bring this message to the attention of a public who for too long have gradually adopted flawed food production and healthcare systems that are on the verge of collapse, threatening the collapse of entire societies as we know them. More specifically, a public and professional dialog on the meaning of nutrition, especially its wholistic properties, is desperately needed, especially in medical schools where nutrition as a science is almost totally ignored.

Keywords: Healthcare; nutrition; medical practice; whole food plant-based; drug therapy; wholism; reductionism

INTRODUCTION

For two millennia, Hippocrates, the "Father of Medicine" (at least in Western medicine discussions), has often been quoted as having said: "Let food be thy medicine and medicine be thy food." Therefore, using the USA system as an example, it seems ironic that although we use his name to honor the ethics in medicine, we mostly ignore his advice on food and its biological expression, nutrition, in the practice of medicine. Medical schools do not teach the science of nutrition. Primary care practitioners, and their specialist colleagues, find it exceedingly difficult if not impossible to get reimbursement for giving nutrition advice to their patients, and among the 28 research institutes and programs at the prestigious US National Institutes of Health (NIH) for example, not one is dedicated to the science of nutrition. There is no doubt that nutrition, as a sophisticated, scholarly science, receives little or no attention in medical institutions. In short, there is a mismatch and, as a consequence, an incalculable loss to all parties. Having lectured quite extensively in public health-oriented communities in Europe, Asia and Australia, I (T Colin Campbell), see virtually no evidence that the dialogue and practices are any better. So-called Western medicine mostly is the norm in European professional medical communities while, in Eastern communities, there is a tendency to believe that the Western model is superior to their more recent past of indigenous health products and practices.

DRUG THERAPY MODEL

One way to discuss this mismatch is to question the present priority given to drug therapy instead of nutrition as the principle means to maintain health and treat disease. Do we have a reliable understanding of the costs and benefits of using, as a priority, the drug therapy model?

On the question of the time and expense for research and development (R&D), the pharmaceutical industry (Anon, 2015) acknowledges that "the process for researching and developing new medicines is growing in difficulty and length" and, further, that "at least ten years" is required for a new medicine to reach the marketplace following its initial discovery. Testing for the safety and efficacy of these drugs in clinical trials is mandatory, of course, and the time required for these trials, "from start to finish ... takes an average of six to seven years," now having become an industry unto itself. In 2013, for example, pharmaceutical companies "sponsored 6,199 trials in the US, involving 1.1 million participants" (Anon, 2015, p. 10).

According to the Tufts Center for the Study of Drug Development (DiMasi, Grabowski, & Hansen, 2016), which, to a "large extent," is funded by the drug industry,[1] the average cost for R&D of a new drug is US$2.6 billion, having increased at an annual rate of 8.5% above general price inflation. Much of this huge investment is the need to test a much larger number of candidate drugs than those that are approved for marketing. Less than 10% of candidate drugs submitted for clinical trial testing make it all the way to the US Food and Drugs Administration (FDA) approval (Carroll, 2014). Writing from a corporate friendly perspective, Lo (2017) caustically asserts that "a general failure rate of 90% [for new drug approval] is bad enough, but for new drugs targeting complex and poorly understood conditions, drug failure is closer to a certainty than a risk."

But this estimate of US$2.6 billion has been seriously questioned. It includes "time" or "opportunity" costs, namely the money committed to research that could have been invested elsewhere, with an assumed return on investment of 10.5%. If this is not counted, the R&D cost per drug drops to US$1.4 billion. It has been argued that because the Tufts Center is largely funded by the pharmaceutical industry, the higher estimate is industry-friendly which supports higher drug prices in the marketplace. Actually, an earlier 2011 systematic review of 13 estimates of R&D costs revealed a 9-fold range, from US$161 million (US$92 million – cash basis, US$161 – capitalized basis) to US$1.8 billion (US$884 million – cash basis, US$1.8 billion – capitalized basis) (Morgan, Grootendorst, Lexchin, Cunningham, & Greyson, 2011).

These estimates, five years apart, in the public's mind, therefore range from US$92 million (cash basis) to US$2.6 billion (capitalized basis), a 28-fold range (16-fold range if comparing each on the same basis) leaving an uneasy impression that it is virtually impossible to have a reliable estimate for the public. Add to this concern the fact that the raw data for these estimates are provided by the industry itself, not from public records that can be examined.

Business Insider, a business news website, also recently claimed, "It's hard to get a clear estimate on how much it costs to develop a new medication from start to finish, with some estimating it costs more that $2 billion." They also said that the average cost "to develop a new cancer drug is closer to $648 million" and, further, that this information "flies in the face of the argument that drug companies need to set high prices to recoup their investments" (Ramsey, 2017).

In addition to the R&D funding, still more funding is required for drug marketing. A 2015 *Washington Post* article (Swanson, 2015) reported that among the top 10 pharmaceutical companies, all but one was spending more on total drug marketing than on R&D. Johnson & Johnson, the biggest marketer, spent US$17.5 billion, more than double their R&D budget.[2] Of the 10 most advertised drugs in 2015, the marketing of a single drug ranged from US$133 million (Viagra) to US$357 million (Humira). About 90% of the marketing budget is for "educating" healthcare professionals than on directly informing consumers, although consumer marketing is very effective. It has been reported[8] that, after seeing a drug or medical device advertisement and being told its side effects, two-thirds of the viewers act on it, with 40% making a follow-up appointment with their doctor. According to a report of the Pew Charitable Trusts, information on side effects may actually increase the credibility of the advertised message because about 43% of viewers believe that the drugs, if advertised, are safe and effective (Kaufman, 2017).

To say that we can know with any confidence how much money is spent on the research, development and marketing of drugs is wishful thinking. We can only say that it is in the billions of dollars for each drug, but the fact that there is so much uncertainty makes it frustratingly difficult to know whether and how it might be improved. That there are additional uncertainties about the health benefits of these drugs makes it even more difficult. How often are people taking drugs simply to "play it safe," possibly doing more harm than good? What do we know, for example, about the placebo effect? In a study on drug treatment of migraine where pain is easily recognized, 50% of the perceived drug effect was a placebo effect, that is, patients given a more positive message showed a more positive response (Kam-Hansen et al., 2014). A placebo effect of 50% is quite substantial and may be more common than generally appreciated (Sheldon & Opie-Moran, 2017). What, too, are the costs of treating unwanted side effects, which exist for virtually all drugs, some effects of which may be permanent (Graedon, 2014)? We know that mortality from drug side effects is the third or fourth leading cause of death, depending on whether doctor's errors associated with the use of drugs are counted (Giacomini et al., 2007; Starfield, 2000). Drug resistance, which accumulates over time, especially for drugs treating pathogenic (Van Katwyk, Grimshaw, Mendelson, Taljaard, & Hoffman, 2017) and neoplastic diseases (Durrant & Morrison, 2017), is yet another property that makes it difficult to estimate how effective or ineffective drugs really are.

Another perspective on this problem of relying on the drug remedy model as the primary means of improving society-wide health are the several reports during the past three to five decades showing that greater use of pharmaceuticals (like the US) does not associate with lower disease mortality (OECD, 2013). On a specific example, consider the use of cytotoxic chemotherapy agents to treat cancer. A 2004 publication presented the findings of a retrospective analysis of the effectiveness of cytotoxic chemotherapy agents used on 22 types of cancer in Australia (72,903 cancers) and the US (72,903 cancers) and found that 5-year survival rates, in the aggregate, were increased only by 2.3% in Australian patients and 2.1% in American patients (Morgan, Ward, & Barton, 2004). Although a minor benefit was observed for a couple cancers (testicular and Hodgkin's lymphoma), these data suggest that, overall, there is little or no benefit resulting from the treatment of this broad group of cancers with cytotoxic chemotherapy, a finding that could hardly be more depressing and devastating. To use the US experience, which generally relies on private capital to advance healthcare solutions and "progress," we can only lament the vast amount of funds wasted and lives lost since the search for "a cure for cancer" was begun over 40 years ago with the War on Cancer begun by the Nixon administration.

It should not be forgotten, however, that there is convincing evidence that chemotherapy is useful in some cases. For example, a series of meta-analyses of 123 studies (100,000 cancer patients) conducted by the Clinical Trial Service Unit at the University of Oxford under the direction of Sir Richard Peto, certainly one of the most reliable and well conducted such study ever undertaken (EBCTCG, 2011), showed that during a 10-year period following treatment of ER-positive breast cancer patients with multiple chemotherapy regimens, the authors reported that the "risk of death from breast cancer can be reduced by about a third." The chemotherapy agents in this analysis belong to the taxane class of chemicals originally isolated from the Pacific Yew tree. Their principle mechanism of action appears to be their ability to disrupt microtubule function essential to cell division – stopping cancer growth.

An estimate of one-third reduction in mortality for this specific cancer by these chemotherapy agents certainly merits their use in the clinic, but the two-thirds of patients who will not benefit should also be considered. Nor should we overlook the cost – health and money – of any unintended side effects caused by these highly toxic agents, like acute myeloid leukemia and cardiotoxicity that have been reported for one of the chemotherapy agents (anthracycline) (Azim, de Azambuja, Colozza, Bines, & Piccart, 2011, Praga et al., 2005). Both effects were observed in some of the studies included in these meta-analyses, albeit at a very low rate, but also the reports for some of the clinics did not record this information, making it difficult to assess the impact of adverse effects on quality of life. Also, it should be noted that none of these women were using the whole food plant-based diet where risk reduction could be nil.

To re-emphasize, determining the costs and benefits of the drug remedy model, which is used as the chief protocol for a nationwide healthcare system, is not achievable. The raw data needed for determining these costs are not publicly available in a form that can be verified and, further, the data which are available are inconsistent and seriously compromised by conflicts of interests. Similarly, the claims of health benefits for the society as a whole are often not adequately qualified by professionals. But one observation is impressive. The total worth of the industry is said to top US$1 trillion (as of 2016) (Statistica, 2017) and, if this amount were to represent the economy of a country, it would be the 16th largest country in the world (Statistics Times, 2017). Although there is convincing evidence for using pharmaceuticals to help resolve many specific illnesses for individuals, there is little or no convincing evidence to our knowledge that relying on drug therapy as the centerpiece for national health policy is a wise choice.

NUTRITION THERAPY MODEL

The limitations of the drug therapy model considered here stand in contrast to an evidence-based nutrition therapy model that prevents and treats sickness more effectively than all the pills and procedures combined and it does so at a cost that is a mere fraction of the cost required for drug therapy. To appreciate this latter point, it is essential that the fundamental biological bases for nutrition and drug therapy (pharmacology) be understood.

The main point of the drug therapy model assumes that there is a key, perhaps rate-limiting biological mechanism responsible for the formation of each disease. Upon identifying this mechanism, a chemical antidote is found that might control, perhaps block, this mechanism. This is the main point of "targeted drug therapy" which, ideally, acts on the target mechanism without damaging nearby healthy cells and tissues. In more recent times, massive efforts are being made to identify the genes responsible for producing the enzymes that catalyze the targeted reactions. A still further extension of this hypothesis suggests that specific mechanisms for disease formation are likely to be unique for each individual person and for each individual disease and/or its variant.

This proposition infers that disease treatment in the future might need to create custom-made drugs and combinations of drugs that are unique for each individual person and for each type of disease, a belief that underlines the popularized concepts of "precision medicine" and "personalized medicine." However, identifying these drugs and making combinations of them is not as easy and logical as it may seem. Within the body, and within the cell, a drug must negotiate a pathway through an infinitely complex maze of countless biochemical mechanisms that, in the normal, healthy state are carefully orchestrated as if they were a symphony. The word often used to establish this state is homeostasis. When chronic, degenerative diseases begin to form, they begin as a distortion of that symphony of mechanisms, some outlying mechanisms of which may be used as predictors or biomarkers of eventual disease as they begin to form. It stands to reason that designing a specific drug to attack a specific target buried within a highly complex milieu of mechanisms very likely will damage neighboring mechanisms (i.e., side effects) on that journey to its target site. Such side effects are commonly observed for virtually all drugs. Add to this the fact that these chemicals almost always are foreign to the body because the natural or parental form of these chemicals cannot be patented as drugs. These foreign chemicals add a further burden to the body which needs to "de-toxify" and excrete these drugs from the body.

Instead of targeting one deranged mechanism in a distorted symphony, would it not make more sense to restore the integrity of this symphony by using a process that acts comprehensively and naturally? Nature had plenty of time to create that amazingly complex symphony of mechanisms to create health, undoubtedly relying on a comprehensive collection of chemicals in food to do so. As the symphony begins to fall into disarray in response to consuming the wrong food, for example, is it not reasonable to assume that Nature would call on those same factors to re-create that symphony? In reality, this is the de facto explanation of nutrition.

The biological basis for nutrition represents a wholistic instead of a reductionist function (Campbell, 2017a). However, this has not been appreciated, even in the professional field of nutritional science. Traditionally, and much to its discredit, research investigations of nutrition have mostly relied on reductionist philosophy that focuses on the effects of single nutrients acting independently, either practically by using single nutrient supplements or theoretically by interpreting food associations with health and disease as associations of nutrients in the food. Foods are valued for their nutrient contents, recommendations of foods to be consumed are initially based on the requirements of individual nutrients studied in isolation and packaged foods (however, they may be processed) are valued for their individual nutrient contents listed on the package.

The focus on the chemical and biological properties of individual nutrients reached a zenith during the 1920s and 1930s when specific nutrients were discovered and alleged to be the causes of so-called nutrient deficiency diseases like beri-beri (thiamin), xerophthalmia (vitamin A), scurvy (vitamin C), rickets (vitamin D) and pellagra (the B vitamin niacin). A second zenith appeared during the 1980s and 1990s with the commercialization of nutrient supplements and the fortification of foods with nutrients. Although research on individual nutrients and their subsequent exploitation as individual nutrient supplements were helpful in elucidating nutrient function, the causation and control of disease have proven to be more complex than simply assuming single nutrient therapy. During the past 20–30 years, many studies have been conducted on the medicinal use of single nutrient supplements for a large number of ailments and diseases and findings have not proven to be effective (Kolata, 2003; Morris & Carson, 2003). Not infrequently, these studies have shown opposite responses for nutrients provided as food and when provided as supplements.

In contrast, whole reasonably intact foods from the plant kingdom are being shown to reverse diseases like coronary heart disease (Esselstyn, Gendy, Doyle, Golubic, & Roizen, 2014; Ornish et al., 1990) and type 2 diabetes (Barnard, Cohen, & Ferdowsian, 2009; McDougall, 2005/11) while substantial evidence has been published during the past few decades to support a similar effect on a broad spectrum of diseases (Expert Panel, 1997; WCRF, 2007). We have now reached a time when the evidence on the ability of whole plant-based foods (WFPB) (vegetables, whole grains, fruits, legumes and nuts) to create and maintain health and reverse disease is more than adequate to maintain health and to prevent and even to treat a broad spectrum of diseases.

The evidence supporting the WFPB diet has gradually evolved over recent decades. Thousands of research publications on selected topics and hundreds of conference papers and institutional reports have shown consistent trends in favor of a whole food plant-based diet. But individuals exclusively using the

WFPB diet itself have not been included in these studies. Vegans and vegetarians are not the same. In the largest single survey of vegans, vegetarians and meat eaters show that consumption of total fat and sugar were almost identical; fat was 30–31% and sugar was 22–23% of total diet calories for all three diets (Sobiecki, Appleby, Bradbury, & Key, 2016). Vegans consume no animal-based foods but are consuming a considerable amount of convenience foods which are high in added fat and sugar. Approximately 90% of vegetarians still consume copious amounts of cow's milk and other dairy products. The WFPB diet, in contrast, ideally includes vegetables, fruits, grains, legumes, and nuts in the whole food form with no added oil, most of which is the omega-6 pro-inflammatory type. The scientific basis for the WFPB diet is more completely described in a recent paper (Campbell, 2017a) and before that, in the extensively referenced books, *The China Study* (Campbell & Campbell, 2005) and *Whole: Rethinking the Science of Nutrition* (Campbell, 2013).

For purposes here, we will summarize the empirical evidence for the WFPB diet from three perspectives: observational, biological plausibility and intervention. On comparing populations, mortality rates for cardiovascular diseases among 20 countries (Jolliffe & Archer, 1959), breast cancer among 38 countries (Carroll, Braden, Bell, & Kalamegham, 1986), "intestinal" cancer among 28 countries (Gregor, Toman, & Prusova, 1969) and colon cancer among 37 countries (Drasar & Irving, 1973) are linearly correlated (r=~0.8) with animal protein consumption. Making this observation particularly notable is that the regression line passes through the X:Y origin, suggesting that there is, on average, no theoretically "safe" level of animal protein consumption. This does not, of course, prove that these mortality rates are specifically caused by animal protein, but animal protein certainly does refer to whole animal-based food. Thus, there are likely many nutrients in a diet rich in animal-based foods that should be participating in the correlation, positive correlations for animal food components (e.g., saturated fat) and inverse correlations for nutrients of plant-based foods (dietary fiber, antioxidants). In other words, a correlation with "animal protein" is, in part, an indication of decreased consumption of whole plant-based foods. In a compilation of findings of 178 observational studies on the association of vegetable consumption with breast cancer risk, of the 144 that showed statistically significant associations, all 144 studies reported an inverse association – increased vegetable consumption and decreased breast cancer mortality (Expert Panel, 1997). Collectively, the consistency and effect size of this evidence support the classical criteria for proof of causation but as in whole food, not as single nutrient (Hill, 1965).

Another criterion for assessing "proof of causation" for the WFPB dietary lifestyle concerns whether it is biologically plausible (Hill, 1965). But in this case, it is necessary to consider the biological plausibility of many nutrients not one. Because dietary associations with cancer and related diseases mainly separate into contrasting biological effects of animal- versus plant-based foods and because dietary choice has long been conditioned by an intense desire for the perceived health benefits of protein-rich animal-based food (Campbell, 2017b, 2017c), the following considers the biological plausibility of animal protein promotion of experimental cancer, specifically the effect of an animal-based protein (casein) on the development of liver cancer initiated in laboratory rats by a powerful carcinogen (aflatoxin). The original hypothesis in this series of studies was intended to identify the key mechanism responsible for the initiation and promotion of casein. When animal-based casein exceeded the amount otherwise needed for normal physiological function (about 10% of total diet calories), it dramatically increased tumor development. Tumor development also could be reversed, either by replacing the animal-based protein with plant-based protein or by decreasing dietary animal protein from 20% to a level less than 10% of total diet calories. The effect size was huge. All animals fed the 20% animal protein-based diet were tumor-bearing but for animals fed the diet containing less than 10% animal protein, none were tumor-bearing.

Identifying the key mechanism responsible for cancer development was not possible. As summarized elsewhere (Campbell, 2017d), each time an hypothesized key mechanism was investigated, it showed the same effect in response to increasing dietary casein. The enzyme responsible for causing more mutations and more cancer initiation was increased, covalent bonding of carcinogen to DNA increased, growth hormone that encourages cell replication increased, calorie metabolism shifted in a way unfavorable to cancer growth (increased brown adipose tissue and increased thermogenesis) and production of reactive oxygen species that (McDougall, 1994) favors cancer promotion increased. Two other mechanisms (DNA repair and natural killer cell activity), however, were decreased by high animal protein feeding but both of these are innate mechanisms that naturally assist the body to block cancer development. Each of these mechanisms, whether increased or decreased by elevated animal protein consumption, resulted in promotion of cancer development. Two observations were apparent. First, multiple mechanisms working together is likely to be far more important than a single key rate-limiting mechanism in explaining causation. Second, this evidence strongly supports the observation that the association of animal protein consumption with cancer is biologically causal.

However, the association of animal-based foods with cancer and other degenerative diseases cannot be specifically and solely attributed to animal protein. There will be an almost unlimited number of other nutrients whose consumption levels will be determined by the consumption of animal-based foods. Most notably, consumption of nutrients of plant-based foods, which are known to prevent cancer, will decrease, resulting in the combined effect of animal and plant-based nutrients that mutually support the same effects on disease formation. Moreover, evidence suggests that the specific effects of individual plant chemicals on cancer causation will similarly employ a multi-mechanistic, wholistic strategy (Campbell, 2017a).

The need to obtain evidence on biological plausibility for a causal effect of whole food on disease outcome goes well beyond the traditional practice of assessing the plausibility of a single nutrient cause-single effect relationships. Moreover, it is necessary to show consistency of effect for the multiple nutrients in the same whole food.

A third demonstration of disease causation by a WFPB diet is the evidence from intervention studies. Coronary heart disease, for example, can be reversed, even if in a relatively advanced stage of development. Its effect size is almost spectacular. Of 197 diagnosed heart disease patients who were adherent to the WFPB diet for a mean of 3.7 years, disease recurrence was 0.6% for those who adhered to the dietary lifestyle, while it was 62% for those who failed to follow the advice (Esselstyn, Ellis, Medendorp, & Crowe, 1995; Esselstyn et al., 2014). A similar slightly earlier finding was reported (Ornish et al., 1990) with additional evidence showing that the WFPB diet also associated with increased telomere length, an indicator of ageing, disease and premature mortality (Ornish et al., 2013).

Evidence is cited here from three kinds of experimental findings (epidemiological, mechanistic and intervention) that demonstrates a relatively rapid, sustainable and broad-based effect of the WFPB diet on the prevention and treatment of complex, degenerative diseases like heart disease and cancer. In addition, this evidence shows that it is a large effect size. A similar dietary effect has been shown for the treatment of diabetes (Barnard et al., 2009; McDougall, 2011) and related illnesses (McDougall, 2002; McDougall, Litzau, Haver, Saunders, & Spiller, 1995). Unfortunately, too many of these observations concerning this dietary effect have gone relatively unnoticed by mainstream media because, in part, they have not been published in the peer-reviewed science and medical journals. Health clinics like those of the Pritikin Center (Hall, Dixson, Barnard, & Pritikin, 1982), the McDougall Program (McDougall, 1994; McDougall & McDougall, 1983), the True North program (Goldhamer, Lisle, Parpia, Anderson, & Campbell, 2001) and the Coronary Health and Improvement Program (Ludington & Diehl, 2000) of the Adventist community all have a track record of more than a quarter century and have demonstrated a quick return to health when people are provided this diet.

We believe that the information presented here on the health benefits of a WFPB diet is sufficiently convincing to prompt the question why is this information not better known? How does the drug therapy model, the centerpiece of our healthcare system, compare with the nutrition therapy model, as exemplified by the WFPB dietary lifestyle? And why is it not recognized? The answer is quite simple. The drug therapy model is far more capital intensive but, as a national policy, provides little or no health benefits whereas the nutrition therapy model is far less capital intensive but provides far more health benefits. The drug therapy model treats specific (targeted) biological mechanisms to control disease (reductionism) whereas the nutrition therapy model treats a symphony of countless mechanisms (wholism) to control disease. By focusing on specific targets, specific chemical antidotes are synthesized and secured with patents in order to insure a grace period of exclusive ownership and a return on investment. By focusing (with a wide lens) on a large, natural scope of related targets collectively involved in health and disease, whole plant-based foods may be used naturally with very little investment. The drug therapy model depends on synthetic chemicals that are foreign to the body, thus generating side effects. The nutrition therapy model depends on whole food for which there is no need or even any possibility for intellectual property protection.

These are the essential differences between the drug and nutrition therapy models and "ne'er shall the twain meet," as the old saying goes. The nutrition therapy model depends on and uses Nature's order of things to good effect. Almost every effort is made by a system beholden to the drug therapy model to minimize or even suppress information and programs that might assist in the development of a nutrition therapy model. Business-wise, it would be a competitive, possibly even a destructive technology. As said previously, physicians are trained to use drugs (reductionism), not food (wholism), to maintain and restore health and physicians get no training in nutrition (certainly not on WFPB nutrition). Original research focused on the health promoting effects of nutrition is severely limited and denigrated and physicians have little or no opportunity to obtain compensation for nutrition counseling. The science of nutrition itself is its own worst enemy because they mostly ignore the wholistic basis for the nutrition effect by mischievously labeling nutrients as "nutriceuticals," then discussing their "chemo-preventive" effects as if these are mere chemical reactions independently working to create the effect. This makes them subservient to the food industry. Make no mistake about it – the divide between these two perspectives on how to advance human health does not lie on a level playing field. The chasm between them is deep and a source of wide differences.

If a national healthcare system is ever to work for the **health** of its citizens, this information needs to be kept in mind in order to know what is at stake thus in order to advance a system that gives a lot more weight to food and nutrition as the means to health than to drugs as the means to health

PUBLIC INTERVENTION MODEL

Given this awareness, the following is a model that we believe has a chance to contribute to this field of public health where economics, science and social justice are jointly used to create a system better than the drug therapy system now used. It should be noted that this new model, in order to get access to the necessary capital for operations, is comprised of for-profit and not-for-profit entities that operate independently but still work together to produce the desired solution. Also, one author (T. Nelson Campbell) has vested interest in this program while the other (T. Colin Campbell) does not.

When we discuss strategies for societal change, the first impulse is to discuss governmental policy. Yet, when we consider history, rarely if ever is transformative change made in a top-down manner, especially in large, complex societies.

Transforming our world around plant-based nutrition is a daunting challenge. There are many governmental, economic and cultural barriers to this change, but the greatest impediment is the problem of personal bias and the difficulty many of us have even seeing these biases. This is especially true with food and health, which is deeply personal to many of us. Overcoming these challenges will require more than can be delivered only through public policy.

This is not to diminish what a government can do to support the change process, most important of which is simply insuring that the public gets to know the objective facts of nutrition and health. Truth-telling can be done by governments, and truth-telling is always the first step in any revolution. This has not happened yet, however, and we would be wasting our time waiting for this to happen. The economic interests benefitting from our poor health account for trillions of dollars of wealth, and it is this economic power that controls public health policy and dissemination of information.

To illustrate this concern, here are two examples in the United States which we know best and for which we (mostly T Colin Campbell) have had first-hand experience. We hesitate to consider government programs of other countries for which we have very little or no personal experience. Also, we believe that the most meaningful information comes from personal experience in the decision-making process rather than relying on third party publications because important public policy decisions are often determined behind closed doors out of public view.

In the US, the Department of Agriculture (USDA) creates and publishes the nation's dietary guidelines, a formal program begun in 1980 and refreshed every five years by a committee of professional experts. A preliminary committee report is then opened for public comment and, after consideration of these comments, the committee then submits their report to the chief USDA administrator, the Secretary of Agriculture, a political appointee of the US President. The report final message therefore is ultimately controlled by a political appointee, who is almost never an expert on the topics of diet, nutrition and health. The committee of scientists, who are very much aware of their limited influence on the final outcome, generally tread lightly in their remarks in their hope to secure the Secretary's approval.

The most recent committee report for the years 2015–2020 secured an unusually large number of public comments (about 30,000), seemingly reflecting a growing interest in the potential impact of the protein-rich American diet on climate change. This interest had been growing in recent dietary guideline reports both in reference to the accumulating evidence on the health value of plant-based foods and on the impact of consuming animal-based foods on the environment (Goodland, 2011; Goodland & Anhang, 2009).

During the public comment period, I (refers to T Colin Campbell) commented on the committee's failure to mention, in their discussion of the perennial diet-heart disease association, recent findings on an unusual ability of the whole food plant-based diet to reverse/cure heart disease (Esselstyn & Golubic, 2014). My comments were selected and published by the US Congress newspaper, *The Hill*, (Campbell, 2015) but even given this attention, neither the committee nor the Secretary acknowledged this important finding in the report. This has been the case for every 5-year report since 1990, when information on the clinical effects of a whole food plant-based diet on heart disease was first published by Ornish and colleagues (Ornish et al., 1990). In passing, it is important to note that the 2002 report updated prior dietary guideline recommendations on dietary protein by increasing the so-called "safe" level of protein to 35% of total calories, an unusually audacious, immoral and dishonest conclusion. Not so incidentally, the majority of the committee's members had or were receiving funding from the dairy industry while the chair of that committee was a well-known consultant to the dairy industry whose livelihood depends on their provision of animal-based protein. He also had failed to adequately disclose the amount of his personal compensation from that industry, as required.

Having been a member of several high-profile expert panels on diet and health, I know well both the practice of scientists to speak softly on this topic and the practice of government authorities to minimize public awareness of important information. I also know the private views on this information by many people in authority but also their reluctance to say the same for the public. Elected officials fear retribution from those who fund their election campaigns. Second and third tier administrators fear loss of their jobs if their superiors fail to be elected.

Another experience illustrating how unusually significant issues on diet and health are decided beyond the public's eyes and ears was my membership on a US National Academy of Sciences expert panel of 13 scientists who wrote the most sought-after report in the Academy's history, the 1982 Diet, Nutrition, and Cancer report (Committee on Diet Nutrition & Cancer, 1982). This was the first institutional report suggesting a national goal of decreasing average dietary fat to 30% of total diet calories as a means of preventing cancer. It chose this specific 30% benchmark and not a lower evidence-based benchmark because this might have inferred decreased consumption of dietary protein from animal sources, an extremely sensitive topic. This National Academies of Science (NAS) report eventually became a landmark publication with several less than desirable outcomes for decades thereafter. The committee's refusal to suggest an evidence-based goal for dietary fat lower than 30% of total calories shows how highly consequential public policy can be shaped out of public eyesight and how scientists, who are responsible for the evidence, have limited influence on national policy on diet and health.

I have had many similar experiences that often reflect an extreme reluctance by government policy-making agencies to tell science as it is. Hot button issues like questioning the alleged health benefits of animal protein-rich diets, the vast overuse of pharmaceutical remedies for human health and the role of food choice on environmental catastrophe are topics unlikely to reach the public anytime s. Political motivations often reign supreme.

Fortunately, we do not need to wait for seriously compromised governments to do the right thing. We live in world today where individuals and groups can tell facts as they see them to large numbers of people through nothing more than key strokes. We also are entering a period where mass local action has never been easier to organize and inspire.

The program described here, PlantPure (mostly involving T. Nelson Campbell), is in the midst of developing a community-based model based on this hopeful view of change (HealingAmericaTogether.com). Importantly, this model is not unique to the US, but has application globally. Transforming the entire world around plant-based nutrition has never been more urgent, especially because of the contribution of animal agriculture to our rapidly accelerating climate change (Goodland & Anhang, 2009; Steinfeld et al., 2006).

This model will utilize various tools, including education, low cost food and an innovative web-based social networking platform to organize, launch and support community-wide health campaigns. These campaigns will bring the message of plant-based nutrition to every sector of a local community, using proactive strategies that educate, provide practical support, and build peer pressure on appropriate institutions to support a personal change that many people who are left alone find hard to make. Importantly, this community model will include a focus on underserved neighborhoods, where the needs are highest.

Leadership for these campaigns will come from individuals and organizations in local communities, including employers, nonprofits, faith-based organizations, and city governments. Initial success will be captured on film and streamed out to the public in the form of short videos, to inspire rapid replication of the model in other communities.

This model also has repercussions beyond health. It has the potential of unifying people of opposing political viewpoints within a new paradigm based on the idea of community empowerment to solve all kinds of problems. The social activism more common on the political Left and the distrust of centralized government more common on the Right are not mutually exclusive. Both can co-exist within a new paradigm focused on the idea that social justice can be achieved best when people are connected, informed, resourced, and inspired to heal the world around them, one person, one family, one neighborhood, and one community at a time.

The world is destined to go plant-based because this health truth is one of the most powerful truths ever identified, with vast potential health, economic, social, political and environmental benefits. The critical question is not whether this will happen, but how quickly. Animal agriculture is the single most important contributor to global warming, which is accelerating as poorly understood feedback loops begin to kick in. Our time is running short on several fronts, but through grassroots strategies which empower people and communities to make change and which stand opposed to top-down hierarchies, we believe that we can still control our collective destiny.

NOTES

1. doi:10.1016/j.jhealeco.2016.01.012 (2016).
2. See also https://www.drugwatch.com/featured/big-pharma-marketing/ (2016).

REFERENCES

Anon. (2015). Biopharmaceutical research and development. The process behind new medicines. *PhRMA*, 24. Obesity and Overweight. Centers for Disease Control. FastStats - Overweight and Prevalence. Retrieved from rd_brochure_022307.pdf

Azim, H. A., Jr., de Azambuja, E., Colozza, M., Bines, J., & Piccart, M. J. (2011). Long-term toxic effects of adjuvant chemotherapy in breast cancer. *Annals of Oncology*, *22*(9), 1939–1947, doi:10.1093/annonc/mdq683.

Barnard, N., Cohen, J., & Ferdowsian, H. (2009). A low-fat vegan diet and a conventional diabetes diet in the treatment of type 2 diabetes: a randomized, controlled, 74-wk clinical trial. *The American Journal of Clinical Nutrition*, *89*(5), 1588S–1596S.

Campbell, T. C. (2013). *Whole. Rethinking the science of nutrition (with H. Jacobson)*. Dallas, TX: BenBella Books.

Campbell, T. C. (2015). (Ed). *Will plant-based dietary recommendations spur meaningful change?* Washington, DC: The Hill, Congress Blog.

Campbell, T. C. (2017a). Nutrition renaissance and public health policy. *Journal of Nutritional Biology*, *3*(1), 124–138, doi:10.1080/01635581.2017.1339094.

Campbell, T. C. (2017b). The past, present, and future of nutrition and cancer: Part 1 – Was a nutritional association acknowledged a century ago? *Nutrition and Cancer*, *69*(5), 811–817, doi:10.1080/01635581.2017.1317823.

Campbell, T. C. (2017c). Nutrition and cancer: an historical perspective–the past, present, and future of nutrition and cancer. Part 2. Misunderstanding and ignoring nutrition. *Nutritional Cancer*, *69*(6), 962–968, doi:10.1080/01635581.2017.1339094. Epub 2017 Jul 25.

Campbell, T. C. (2017d, Oct 3). Cancer prevention and treatment by wholistic nutrition. *Journal of Natural Sciences*, *3*(10), 1–29.

Campbell, T. C., & Campbell, II, T. M. (2005). *The China study, startling implications for diet, weight loss, and long-term health*. Dallas, TX: BenBella Books.

Carroll, A. E. (2014). *The New York Times* (Online). Nov 18, 2014.

Carroll, K. K., Braden, L. M., Bell, J. A., & Kalamegham, R. (1986). Fat and cancer. *Cancer*, *58*(8 Suppl), 1818–1825.

Committee on Diet Nutrition and Cancer. (1982). *Diet, Nutrition and Cancer*. Washington, DC: National Academy Press.

DiMasi, J. A., Grabowski, H. G., & Hansen, R. W. (2016). Innovation in the pharmaceutical industry: New estimates of R&D costs. *Journal of Health Economics*, *47*(2), 20–33.

Drasar, B. S., & Irving, D. (1973). Environmental factors and cancer of the colon and breast. *British Journal of Cancer*, *27*, 167–172.

Durrant, D. E., & Morrison, D. K. (2017). Targeting the Raf kinases in human cancer: the Raf dimer dilemma. *British Journal of Cancer*, *1*, 1–3.

Early Breast Cancer Trialists' Collaborative Group (EBCTCG). (2011). Comparison between different polychemotherapy regimens for early breast cancer: meta-analyses of long-term outcome among 100 000 women in 123 randomised trials. *Lancet*, *379*, 432–444.

Esselstyn, C. B., Ellis, S. G., Medendorp, S. V., & Crowe, T. D. (1995). A strategy to arrest and reverse coronary artery disease: a 5-year longitudinal study of a single physician's practice. *The Journal of Family Practice*, *41*(6), 560–568.

Esselstyn, C. B. J., Gendy, G., Doyle, J., Golubic, M., & Roizen, M. F. (2014). A way to reverse CAD? *The Journal of Family Practice*, *63*(7), 356–364b.

Esselstyn, C., & Golubic, M. (2014). The nutritional reversal of cardiovascular disease, Fact or Fiction? Three case reports. *Experimental & Clinical Cardiology*, *20*(7), 1901–1908.

Expert Panel. (1997). *Food, nutrition and the prevention of cancer, a global perspective*. Washington, DC: American Institute for Cancer Research/World Cancer Research Fund.

Giacomini, K., Krauss, R. M., Roden, D. M., Eichelbaum, M., Hayden, M. R., & Nakamura, Y. (2007). When good drugs go bad. *Nature*, *446*(7139), 975–977.

Goldhamer, A., Lisle, D. L., Parpia, B., Anderson, S. V., & Campbell, T. C. (2001). Medically supervised water-only fasting in the treatment of hypertension. *Journal of Manipulative and Physiological Therapeutics*, *24*(5), 335–339.

Goodland, R. (2011). *Our choices to overcome the climate crisis*. NGO Global Forum 14, Gwangju, Korea. Retrieved from http://www.hlrn.org/img/documents/2011

Goodland, R., & Anhang, J. (2009). Livestock and climate change. *World Watch*, 1–10.

Graedon, T. (2014). How long do drug side effects last? For some, perhaps forever. *The People's Pharmacy*. Retrieved from www.peoplespharmacy.com/about

Gregor, O., Toman, R., & Prusova, F. (1969). Gastrointestinal cancer and nutrition. *Gut*, *10*, 1031–1034.

Hall, J. A., Dixson, G. H., Barnard, R. J., & Pritikin, N. (1982). Effects of Diet and Exercise on Peripheral Vascular Disease. *The Physician and Sportsmedicine*, *10*, 90–101, doi:10.1080/00913847.1982.11947226.

Hill, S. A. B. (1965). The environment and disease: Association or causation? *Proceedings of the Royal Society of Medicine*, *xxx*, 295–300.

Jolliffe, N., & Archer, M. (1959). Statistical associations between international coronary heart disease death rates and certain environmental factors. *Journal of Chronic Diseases*, *9*, 636–652.

Kam-Hansen, S. et al. (2014). Altered placebo and drug labeling changes the outcome of episodic migraine attacks. *Science Translational Medicine*, *6*, 218ra215, doi:10.1126/scitranslmed.3006175.

Kaufman, J. (2017). In *The New York Times* (Online). Dec 24, 2017.

Kolata, G. (2003). Vitamins: More may be too many (Science Section). *The New York Times* 1, 6.

Lo, C. (2017). Counting the cost of failure in drug development. (2017). Counting the cost of failure in drug development – Pharmaceutical Technology. June 19, 2017.

Ludington, A., & Diehl, H. (2000). *Health power. Health by choice not be chance* (251 p.). Hagerstown, MD: Review and Herald Publishing Association.

McDougall, J. (2002). Diet: Only hope for arthritis. *McDougall Newsletter*. Santa Rosa, CA: Dr. McDougall's Health & Medical Center. Retrieved from https://www.drmcdougall.com

McDougall, J. (2005/11). *Series of 10-day trials to reduce diabetes symptoms*. Santa Rosa, CA: McDougall's Health & Medical Center. Retrieved from https://www.drmcdougall.com

McDougall, J. A. (1994). *The McDougall program for maximum weight loss*. New York, NY: Penguin Books.

McDougall, J. A., & McDougall, M. A. (1983). *The McDougall plan*. El Monte, CA: New Win Publishing.

McDougall, J., Litzau, K., Haver, E., Saunders, V., & Spiller, G. A. (1995). Rapid reduction of serum cholesterol and blood pressure by a twelve-day, very low fat, strictly vegetarian diet. *The Journal of the American College of Nutrition*, *14*, 491–496.

Morgan, G., Ward, R., & Barton, M. (2004). The contribution of cytotoxic chemotherapy to 5-year survival in adult malignancies. *Clinical Oncology: A Journal of the Royal College of Radiologists*, *16*, 549–560.

Morgan, S., Grootendorst, P., Lexchin, J., Cunningham, C., & Greyson, D. (2011). The cost of drug development: a systematic review. *Health Policy*, *100*, 4–17, doi:10.1016/j.healthpol.2010.12.002 (2011).

Morris, C. D., & Carson, S. (2003). Routine vitamin supplementation to prevent cardiovascular disease: a summary of the evidence for the U.S. Preventive Services Task Force. *Annals of Internal Medicine*, *139*, 56–70.

OECD. (2013). Health at a glance 2013. OECD indicators.

Ornish, D. et al. (1990). Can lifestyle changes reverse coronary heart disease? *Lancet*, *336*, 129–133.

Ornish, D. et al. (2013). Effect of comprehensive lifestyle changes on telomerase activity and telomere length in men with biopsy-proven low-risk prostate cancer: 5-year follow-up of a descriptive pilot study. *The Lancet Oncology*, *14*, 1112–1120.

Praga, C. et al. (2005). Risk of acute myeloid leukemia and myelodysplastic syndrome in trials of adjuvant epirubicin for early breast cancer: correlation with doses of epirubicin and cyclophosphamide. *Journal of Clinical Oncology*, *23*, 4179–4191. doi:10.1200/JCO.2005.05.029.

Ramsey, L. (2017). A new study undermines Big Pharma's justification for charging high drug pricies. *Business Insider*. Retrieved from http://www.businessinsider.com/drug-costs-to-develop-a-new-cancer-drug-2017-9

Sheldon, R., & Opie-Moran, M. (2017). The Placebo Effect in Cardiology: Understanding and Using It. *Canadian Journal of Cardiology*, *33*, 1535–1542. doi:10.1016/j.cjca.2017.09.017.

Sobiecki, J. G., Appleby, P. N., Bradbury, K. E., & Key, T. J. (2016). High compliance with dietary recommendations in a cohort of meat eaters, fish eaters, vegetarians, and vegans: results from the European Prospective Investigation into Cancer and Nutrition-Oxford study. *Nutrition Research*, *36*, 464–477. doi:10.1016/j.nutres.2015.12.016.

Starfield, B. (2000). Is US health really the best in the world? *JAMA*, *284*, 483–485.

Statistica. (2017). The Statistics Portal. Retrieved from https://www.statista.com/topics/1764/global-pharmaceutical-industry/

Statistics Times. (2017). *Listed of countries by projected GDP*. List of countries by projected GDP – StatisticsTimes.com. Retrieved from https://www.scribd.com/document/368710297/List-of-Countries-by-Projected-GDP-2017

Steinfeld, H., Gerber, P., Wassenaar, T., Castel, V., Rosales, M., & de Haan, C. (2006). *Livestock's long shadow: environmental issues and options*. Rome: FAO.

Swanson, A. (2015). Big pharmaceutical companies are spending far more on marketing than research. Washington DC: Washington Post. Retrieved from https://www.washingtonpost.com/news/wonk/wp/2015/02/11/big-pharmaceutical-companies-are-spending-far-more-on-marketing-than-research/?utm_term=.19c58c1076d8

Van Katwyk, R., Grimshaw, J. M., Mendelson, M., Taljaard, M., & Hoffman, S. J. (2017). Government policy intervention to reduce human antimicrobial use: protocol for a systematic review and meta-analysis. *Systematic Reviews*, *6*, 256, doi:10.1186/s13643-017-0640-2.

World Cancer Research Fund (WCRF)/American Institute for Cancer Research. (2007). *Food, Nutrition, Physical Activity, and Prevention of Cancer: A Global Perspective*. Washington, DC: American Institute for Cancer Research.

INDEX

Academy of Nutrition and Dietetics, 7
Accountability, 25, 66, 71, 80, 88, 104, 108
Adaptability, 74, 109
Advanced consent, 158, 161, 163
Advanced Therapy Medicinal Products, 107, 112*n*12
Advocacy groups, 17, 79
Allocational vulnerability, 156
Alzheimer's Association, 160
American Chemical Society, 95
American Society for Parenteral and Enteral Nutrition, 7
Animal research, 2–3, 8*n*4, 24, 29*n*15
AnimalTestInfo, 3
Animal welfare, 2, 14, 116
Anonymized data/anonymization, 45, 47, 48, 55–58, 68, 69, 74, 78, 79
Antiepileptic drug (AEDs), 75–76, 77
Arms Trade Treaty (2014), 139
Association for Responsible Research and Innovation in Genome Editing (ARRIGE), 6, 115–125
Association of Internet Researchers Ethics Working Committee, 58
AstraZeneca, 65
Australia, 123, 186
  diet therapy, 174
  drug therapy model, 189
Autonomy, 24, 29*n*14, 66, 68–70, 78, 79, 102, 103, 110, 162
  and capacity to consent, 153
  ethical principle of, 149–153
  and informed consent, 150–153
Beauchamp & Childress, 24, 29*n*14, 149
Being in public, 51
Being public, 51
Beneficence, 24, 162
Benefit, as criteria for research, 72
Berkelely University, 119
Bias, 45, 47, 54, 67, 77, 120, 149, 195
Big Data, 45–47, 49–54
  bias and representativeness of research, 54
  claims against data sharing, 75
  core values, attachment of, 67–75
  data use, reuse, and misuse, 50–53
  defined, 64–66
  in genetic data, 77–79
  in healthcare and life sciences, 63–80
  for health professionals and researchers, 64–66
  in health research, 77
  online recruitment, 53
  in public health, 76–77
  research ethics in digital era, 54–56, 58
  use in life sciences, 5
Bill & Melinda Gates Foundation (BMGF), 16
Biobanking and Biomolecular Resources Research Infrastructure – European Research Infrastructure Consortium (BBMRI-ERIC), 60*n*20, 65
  Code of Conduct for Health Research, 58
Bioethics, 14, 58, 69, 80, 155
Bioinformatics, 63, 65, 76

Biologic, Toxin and Weapons Convention (BTWC), 139
Biosafety, 87, 88, 90, 93, 118
Biosafety Level 3 (BSL 3), 90
Biosafety Level 4 (BSL 4), 90
Biotechnology and Biological Sciences Research Council (BBSRC), 3
Body mass index (BMI), 169, 178–179
Broad consent, 48, 49, 69–71, 79, 80
BROAD Institute-MIT, 119
Burtscher, Wolfgang, 28
Business Insider, 187–188

Cambridge Analytica, 52
Canadian Animal Care Committee, 3
Capacity to consent
  autonomy and, 153
  informed consent in persons without, 159–162
CBRN Security Culture, 95
Center for Infectious Diseases Research and Policy (CIDRAP), 36
Centers for Disease Control & Prevention (CDC), 35, 36, 39
Cervical cancer screening, in India, 15–16
Charlson Index, 178
Charpentier, Emmanuelle, 119
Charter of Fundamental Rights of the European Union, 47, 59*n*3
  art. 1 (rights to human dignity), 47
  art. 7 (respect of private and family life), 47
  art. 8 (data protection), 47
  art. 13 (right to freedom of arts and sciences), 47
  art. 21 (non-discrimination), 47
Chemical Weapons Convention (CWC), 139
Child health data, 52
China
  genome editing, 117, 123
  international collaborative genetic research project, 16–17
  Kadoorie Biobank, 65
Chinese Academy of Sciences, 111*n*5
Citizenry, 66
Civil security, 90, 91, 100
Clinical dementia research, 7, 147–164
  ethical challenges and issues in, 158–159
    informed consent in persons without capacity to consent, 159–162
  ethical framework and decision-making in
    autonomy, 149–153
    decision-making capacity and competence, 153–155
    vulnerability, 155–157
  ethical questions, 162–163
  recommendations for, 163–164
Clinical Trials Regulation, 103
Cognitive vulnerability, 156
College of Commissioners, 101
Comfort feeding, 7
Commission for the Study of Ethical Problems in Medicine & Biomedical and Behavioral Research, 79
Competence, 100, 103, 104, 150, 153–155, 162
"Complete Khoisan and Bantu Genomes from southern Africa", 17
Convention for the Protection of Human Rights, 59*n*2
  Article, 13, 122
Coronary Health and Improvement Program, 194
Corrective fairness, 25
Council of Europe (CoE), 57, 60*n*17, 111*n*6

Convention for the Protection of Human Rights and Fundamental Freedoms, 59*n*2
Council of State, France, 106, 121
Court of Justice of the European Union, 106, 121
Crick Institute, 117
CRISPR (Clustered Regularly Interspaced Short Palindromic Repeat) array, 106, 107, 115–122, 124
CRISPR-Cas9 system, 90, 99, 100, 102–108, 117, 120, 122, 139
Cultural sensitivity, 23
peer educators for sex workers, 22
safety and ethics, 86–87
Culture of safety and security, shaping, 85–95
added value of, 94
current and future direction of, 93–84
ethics review to safety governance in emerging technologies, using, 89–90
ethics review to security governance in emerging technologies, using, 92–93
Framework Program, 9, 94–95
safety governance, 87–89
security and ethics, 90–91
security governance, 91

DASH (Dietary Approaches to Stop Hypertension) Diet, 172, 173, 176–178
Data
anonymized/anonymization, 45, 47, 48, 55–58, 68, 69, 74, 78, 79
bias in, 45, 47, 54, 67, 77, 120, 149, 195
Big Data. *See* Big Data
donors, data sharing risks for, 75
ethics approach toward, 56–59
extraction, 64
minimization, 48, 49, 59, 75
misuse, 45, 50–53, 57, 58, 67, 80
ownership, 13, 66–67, 76, 77, 79
producers, costs and burdens for, 75
quality, 49, 67, 68, 76
reuse, 50–53
science, 46, 49, 54, 58, 59
security, 48, 55
storage, 66–67
use, 50–53
Data protection, 4, 5, 45–51, 54–57, 60*n*8, 67, 68, 80
regulation of, 47–49
right to, 54, 59*n*4
Data Protection Directive 95/46/EC, 5, 56
Data Protection Working Party Article, 29, 48, 49, 59*n*6, 60*n*7, 60*n*8
Data sharing, 46, 47, 58, 66–67, 71, 72, 74, 76, 77, 80
claims against, 75
Death computer, 77
Decision-making capacity, 7, 147, 149–155, 157, 158, 160
myths of, 155
Declaration of Helsinki, 132
Article, 21, 25
Defense Advanced Research Projects Agency (DARPA), 135
Deferential vulnerability, 156
De-identification. *See* Anonymized data/anonymization
Deliberate Release Directive, 104
Department of Health and Human Services, 16
Devinsky Antiepileptic Drug Study, 77
Dietary Approaches to Stop Hypertension (DASH). *See* DASH (Dietary Approaches to Stop Hypertension) Diet

Diet therapy, 7, 169–181
complexity of, 178–181
defined, 170–171
and science, 175–177
simplicity of, 178–181
Digital era, ethical challenges in, 5, 45–60, 70
Big Data, 49–54
privacy and data protection safeguards, 47–49
research ethics, 54–59
Directive 90/220/EEC, 111*n*8
Directive 90/385/EEC, 112*n*10
Directive 98/44/EC, 112*n*12
Directive 2001/18/EC, 106, 111*n*1, 111*n*8, 111*n*9, 121
Article 2(2), 105
Article, 3, 105
Directive 2001/20/EC, 59*n*5, 86, 111*n*7
Directive 2001/83/EC, 111*n*10
Direct-to-consumer (DTC)
products, 133
testing, 78
Distributive justice, 72–74
Do-It-Yourself (DIY) movement, 121, 133–134, 139
Doudna, Jennifer, 116, 119
DowDuPont Pioneer, 117, 125*n*3
Drug therapy model, for nutrition mismatch, 187–190
Dual use, 6, 92, 103, 108, 109, 112*n*17
definitions of, 130–132
in neuroscientific and neurotechnological research, 129–142
do-it-yourself/neurobiohacking, 133–134
ethical address and guidance, 139–141
extant treaties and regulations, 138–139
military/warfare, 134–138
viability of, 130
Dual Use Research of Concern (DURC), 130–132, 134, 138–140
in neuroscience and neurotechnology, 132–133
Dynamic consent, 70
advantages of, 70–71

Early Breast Cancer Trialists' Collaborative Group (EBCTCG), 189
Ebola hemorrhagic fever. *See* Ebola virus disease (EVD)
Ebola virus disease (EVD), 46
background of, 35–36
care in resource-limited settings, 5, 18, 33–42, 36–38
epidemic, ethical considerations associated with, 42
healthcare workers, fear and stigmatization of, 38–39
repatriation of international workers, from EVD-affected countries, 39–41
EFSA Panel on GMOs, 101, 111*n*3
Electronic health records, 64, 66
Elwa Public Hospital, 40
Enteral nutrition, 7
Equality
gene editing, 102, 110
Equity, 66, 72, 73, 103, 108, 110
promotion, in international research, 11–29
Ethics
approach toward data, 56–59
bioethics, 14, 58, 69, 80, 155
codes, for life sciences, 4
dumping. *See* Ethics dumping
governance, 15
in situ, 6–8
neuroethics, 135, 141
research, 54–59
review. *See* Ethics review
safety and, 86–87
security and, 90–91
Ethics dumping, 11–18, 23, 26, 27, 28*n*1, 90
case studies, 14–18

China, international collaborative genetic research project in, 16–17
India, cervical cancer screening in, 15–16
In life sciences, 4–5
retrospective approval for study in Liberia, seeking, 18
San people-involved international genomics research, 17
tools, 13
Ethics review, 4, 6, 23, 55, 56, 58, 85
to safety governance in emerging technologies, using, 89–90
to security governance in emerging technologies, using, 92–93
EU Law, gene editing governance in, 103–105
Eurobarometer survey, 2, 47
European Academies Science Advisory Council (EASAC), 101
European Commission (EC), 4, 12, 60*n*8, 60*n*15, 90, 131
General Data Protection Regulation (GDPR) 2016/679, 5
Horizon 2020 Ethics Appraisal Procedure, 86, 89, 92–93, 102, 104
Joint Research Centre, 101
Scientific Committees, 101–102
Society Unit, 28*n*1
European Commission Report, 131
European Community, 132
European Convention of Human Rights, 47
European Group on Ethics (EGE), 103, 104
European Medicines Agency (EMA), 106, 132
European Research Area (ERA), 5
European Steering Committee, 6, 123
European Union (EU), 2, 3, 6, 12, 47, 60*n*15, 88, 90
Charter of Fundamental Rights, 47
gene editing governance, 99–112
Exploitation prevention, in international research, 11–29
Exploratory analysis of data, 64

Facebook, 46, 51–53
Fairness, 12, 24–25, 27, 49, 73
corrective, 24
in exchange, 24
Global Code of Conduct for Research in Resource-Poor Settings, 27
Fertility Diet, 172, 174
Flexitarian Diet, 172, 173
Food and Agriculture Organization of the United Nations (FAO), 7
Food and Drugs Administration (FDA), 187
Food and Feed legislation, 103, 111*n*9
4Ps principle, 75
Framework Program 9 (FP9), 94–95
Funder Platform, 18, 19
Fuzziness, 178–180

Gene drive, 99, 101, 102, 107–109, 118, 119, 124
Gene editing, 6, 90, 93, 139
advances of, 116
developments, marketable products derived from, 119–120
extended applications of, 118–119
governance, in EU, 99–112
ethical challenges to, 102–104
EU-wide regulatory intervention, need for, 109–111
gene drive, 107–109
regulatory questions and legal status of, 104–107
responsible use of, 115–125
risks and regulation of, 121–122

General Data Protection Regulation (GDPR) 2016/679, 5, 46–49, 53, 56–57
  art. 4.11, 48
  art. 7.3, 50
  art. 14.5 (b), 49
  art. 31 (1), 57
  art. 35 (1), 57
  art., 89, 49, 55
  art. 3.1ii, 60*n*17
  broad consent, 48
Gene sequencing, 78
Genetically modified organisms (GMOs), 87, 100, 101, 104–106, 109, 111*n*1
Genetic data, Bid Data in, 77–79
Genetic discrimination, 78, 79, 80*n*1
Genetic exceptionalism, 79, 80
Genetic Information Non Discrimination Act (GINA), 78
Genetic privacy, 78, 79
Geneva Protocol, 139
Genome-wide association studies (GWAS), 67
Genomics England, 65
German Ethics Council, 103
Germany
  dual use in neuroscientific and neurotechnological research, 141
  ethical review to safety governance, 90
  repatriation of international workers, 40
Global Alliance for Genomics and Health (GA4GH) for, 65
Global Code of Conduct for Research in Resource-Poor Settings, 5
  Article, 1, 27
  developing, 26–28
  ethics dumping, 27
  Exploitation Risk Table, 27
Google, 46
Hague Convention(s), 139
Harvard Catalyst Regulatory Foundations
  social media research, peculiarities of, 57
Healthcare, Big Data in, 63–80
Healthcare Risk Management, 40
Healthcare workers (HCWs), 15, 37
  Ebola transmission, fear and stigmatization of, 33–34, 38–39, 41, 42
Health research, 15, 46, 58, 63, 64, 66, 76
  Big Data in, 77
High-income countries (HICs), 12, 13, 17, 28
High Level Group of the Commission's Scientific Advice Mechanism (SAM), 55, 57, 101
Hinxton Group, 101
Hippocratic Oath, 180
Honesty, 24–26, 74–75
  intellectual, 74
  radical, 74
Horizon Europe, 12
Horizon 2020 Ethics Appraisal Procedure, 12, 28, 28*n*2, 59, 60*n*21, 86, 89, 92–93, 102, 104
Horizon scanning, 89
House of Commons Science & Technology Committee, 51
Human Fertilisation and Embryology Authority (HFEA), 117
Human Genome Project, 77
Human Heredity and Health in Africa (H3Africa), 65
Human–machine interactive networks, 136
Human rights, 25, 46, 51, 56, 59, 86, 90, 92
Human security, 91, 92

IBM X-Force Research (2017), 53
ICED Index, 178
India, 123
 cervical cancer screening in, 15–16
 ethics dumping, 15
Inductive reasoning, 64
Information disclosure, 151–152
Informed consent (IC), 7, 13, 14, 16, 17, 23, 26, 51, 52, 76, 78–80, 132
 autonomy and, 150–153
 basic elements of, 150–151
 clinical dementia research, 147, 149, 158, 159, 163
 decision-making capacity, 154
 peer educators for sex workers, 21–22
 in persons without capacity to consent, 159–162
  advanced consent, 161
  proxy consent, 160–161
  supported decision-making, 161–162
 respect for, 68–71
 vulnerability, 156, 157
INSERM (the French National Institute for Health and Medical Research) Ethics Committee (IEC), 123
In situ ethics, 6–8
Institute of Electrical and Electronics Engineers (IEEE)
 Model Process for Addressing Ethical Concerns during System Design, 58–59
Institute of Medicine, 71
Institutional Review Boards (IRBs), 18, 132, 159
Intellectual honesty, 74
Intelligence Advanced Research Projects Activity (IARPA), 135
Intensified data sourcing, 64
International Agency for Research on Cancer (IARC), 16
International Association of Gerontology and Geriatrics (IAGG), 148
International Atomic Energy Agency (IAEA), 94, 95
International Declaration on Human Genetic Data (IDHGD), 79
International Ebola Response Team, 38
International justice, 28
International research, equity promotion and exploitation prevention in, 11–29
 ethics dumping, 13–18
 Global Code of Conduct for Research in Resource-Poor Settings, developing, 26–28
 TRUST's global engagement activities, 18–24
Internet. *See* Digital era, ethical challenges in
Interventional research, 71
Intrexon, 118
Investopedia, 13
In vitro fertilization (IVF), 117
Irish Council for Bioethics, 55–56

Johnson & Johnson, 188
Justice, 24, 28, 66, 102, 103, 110, 159, 162
 distributive, 72–74
 gene editing, 102
 social, 195, 198

Kadoorie Biobank, 65
Kenyan Research Ethics Committee (REC)
 research governance *versus* research ethics challenges, 23
 TRUST's global engagement activities, perspective of, 18–24
Ketogenic diet, 175, 176

Lagos Mainland Hospital, Yaba, 39
Legal/regulatory compliance approach, 87
Liberia
  Ebola virus disease, 33–36, 40
  retrospective approval for ethics dumping study, seeking, 18
Life sciences, 3–8, 130
  Big Data in, 5, 63–80
  ethics codes for, 4
  ethics dumping. *See* Ethics dumping
  research production process in, 1–2
Linguistic variables, 169, 178, 179
Lisbon Treaty, 139
Low- and middle-income countries (LMICs), 12, 13, 23, 27, 28, 28*n*9

Malnutrition problems, 7
Mayo Clinic Diet, 172, 173
McDougall Program, 194
Médecins Sans Frontières (MSF), 36
Medical practice, 7, 132, 185–198
Medical research, 24, 25
  ethical challenges in digital era, 5, 45–60
Medical Research Council (MRC), 3
Medical vulnerability, 156
Mediterranean diet, 172–173
Military/warfare dual-use applications, of neuroscientific and neurotechnological research, 134–138
  direct weapinization, 137–138
  human–machine interactive networks, 136
  neural systems modeling, 136
  performance optimization, 136–137
  resiliene in combat and military support personnel, 136–137
Mind control, 138
MIND Diet, 172, 173
Mind reading, 138
Misuse of data, 50–53, 57, 58
Model Process for Addressing Ethical Concerns during System Design, 58–59
Mojica, Francisco Martínez, 115
Morris, Martha Clare, 173
Mortality rate, of EVD epidemics, 37

National Academy of Medicine, 111*n*5
National Academy of Sciences (NAS), 100, 111*n*5, 197
National Action Plan on Breast Cancer (NAPBC)
  Working Group, 79
National Institutes of Health (NIH, USA), 16, 55, 171, 179, 186
  Office of Science Policy (OSP), 130–131
National Research Council of the United States Academy of Sciences (NRC), 134, 135
National security, 90–92, 131, 135, 140
Netherlands, the
  ethical review to security governance, 93
  ethical review to safety governance, 90
  gene editing, 106, 121
  repatriation of international workers, 39, 40–41
Neural systems modeling, 136
Neurobiohacking, 133–134, 140
Neurohacking, 6
Neuroscientific research, dual use in, 129–142
  military/warfare. *See* Military/warfare dual-use applications, of neuroscientific and neurotechnological research

Neurotechnological research, dual use in, 129–142
military/warfare. *See* Military/warfare dual-use applications, of neuroscientific and neurotechnological research
Neurotechnologies Industry Report, 134
New Breeding Techniques (NBT), 101, 110, 121
New Zealand
"Predator-free 2050" plan, 119
War on Rats, 125*n*7
Next-generation sequencing technologies, 67, 78
Niakan, Kathy, 117
Non-discrimination, 47, 56, 57, 59, 110
Non-Human Primates (NHP), 3
Non-invasive brain stimulation (NIBS), 131
Non-maleficence, 24, 102, 130
Nuffield Council on Bioethics, 103
Nuffield Council Report, 134
Nutrition mismatch, medical practice for, 185–198
drug therapy model, 187–190
nutrition therapy model, 190–195
public intervention model, 195–198
Nutrition therapy model, 190–195

Occupational Safety and Health Assessment (OSHA), 87, 89
Office of Science Policy (OSP), 130–131
Off-label use, 131, 132–133
OkCupid, 51
Online recruitment, 45, 53, 54
Open communication, 26
Open data, 45, 52, 60*n*12
Open Data Institute, 60*n*12
Organization for Economic Co-operation and Development (OECD), 47, 67, 188
Ornish Diet, 172, 174

Partners for Health and Development in Africa (PHDA), 18
Sex Workers Outreach Programme, 21
Personalized medicine, 190
Peto, Sir Richard, 189
Pew Charitable Trusts, 188
PlantPure, 197
Population biobanks, 67
Porteus, Matt, 120
Precautionary Principle, 88–89, 90, 100, 122
Precision medicine, 65, 68, 190
Precision Medicine Initiative, 65
President's Commission on Bioethics Report, 79
Principles of Humane Experimental Technique, The (Russell and Burch), 3
Pritikin Center, 194
Privacy, 2, 5, 45, 47–57, 59, 64, 66–69, 71, 75, 76, 78–80, 110, 149, 158
genetic, 78
losses, 72
right to, 56
Proceduralism, 110
Productivity output, 1
Proxy consent, 160–161
Psychological operations (PSYOPS), 137
Public data, 47, 51–52, 57–59
Public health, Big Data in, 76–77
Public health-oriented ethics, 69
Public intervention model, for nutrition mismatch, 195–198

Radical honesty, 74
Reciprocity, 66, 73, 74
Reductionism, 194, 195
Regulation on Medical Devices, 103
Repatriation of international workers, from EVD-affected countries, 39–41

Research and development (R&D), 2, 187–188
Research and innovation (R&I), 2
Research ethics, in digital era, 54–59
Research Ethics Committees (REC), 58, 89, 90, 93, 132, 159
Research integrity, 2, 26, 75
Research production process, in life sciences, 1–2
Resilience, 74, 75, 136–137
Resource-limited countries, EVD care in, 36–38
Respect for autonomy, 149
Respect for informed consent, 68–71
Respect for persons, 68–71, 150, 162
Respect of private and family life, 47
Responsible Research and Innovation (RRI), 86, 94
Restoring Active Memory (RAM), 135
Reuse of data, 50–53
Right to data protection, 59*n*4
Right to freedom of arts and sciences, 47
Right to human dignity, 47
Right to privacy, r6
Risk assessment, 87, 89, 101, 102, 107–109, 141
Risk of harm, 68, 153, 157
Riyadh Intensive Care Program Mortality Prediction Algorithm, 77
Robustness, 74, 75, 118
Royal Society Working Group on Neuroscience, 111*n*5, 141

Safety, and ethics, 86–87
Safety culture, defined, 94
Safety governance, 87–89, 93
  in emerging technologies, ethics review to, 89–90
San Code of Research Ethics, 21, 24
San Diego-based biotech Human Longevity, 65
Sanger Institute, 65
San, the, 18
  people, international genomics research using, 17
  Research and Media Contract, 21
  sex workers, peer educators for, 19, 21–22
Scientific Committee on Consumer Safety (SCCS), 102, 111*n*4
Scientific Committee on Emerging and Newly Identified Health Risks (SCENIHR), 102, 111*n*4
Scientific Committee on Health and Environmental Risks (SCHER), 102, 111*n*4
Scientific value, 67–68
Second International Conference on Nutrition (ICN2), 7
Security, 2, 6, 48, 53, 55, 57, 67, 68, 85–95, 132–134, 139
  civil, 90, 91, 100
  communities, 109
  and ethics, 90–91
  human, 91, 92
  national, 90–92, 131, 135, 140
Security governance, 85, 86, 90, 91
  in emerging technologies, ethics review to, 92–93
Self-governance, 87, 88
Sex workers, 18
  in LMICs, 28*n*9
  in San, peer educators for, 19, 21–22
Sex Workers Outreach Programme, 21
Situational vulnerability, 156
Social vulnerability, 156
Solidarity, 39, 66, 73–74, 110
South African San Institute (SASI), 18
Spain
  repatriation of international workers, 39, 40
Stakeholder engagement, 19
SUBNETS, 135

Supported decision-making, 158, 161–162
Sustainability, 25, 72–73, 108
Sweden
  gene editing, 106, 121
Systems medicine, 63, 65, 76

Therapeutic Lifestyle Changes (TLC) Diet, 172, 173
3Rs principle, 3, 29*n*15
Transcription Activator-Like Effector Nucleases (TALENs), 106, 116
Transparency, 26, 49, 60*n*7, 66, 71–73, 76, 101, 110, 123
Treaty on the Functioning of the European Union
  Article 16.1, 59*n*4
True North program, 194
Trust, 22, 50, 57, 58, 68, 70–72, 77, 80, 99, 110, 111, 138, 180
TRUST project, 4, 5, 11–29
  care, 25
  fairness, 24–25
  global engagement activities
    Kenyan Research Ethics Committee perspective, 23
    meeting of many minds, 24
    sex workers in San, peer educators for, 19, 21–22
  honesty, 25–26
  respect, 25
  timeline of, 20
Tufts Center for the Study of Drug Development, 187
Tutu, Desmond, 17
Twitter, 46, 51, 76

Uncertainty, in safety governance, 88
UNESCO, 103, 104, 110
  International Declaration on Human Genetic Data (IDHGD), 79
United Kingdom (UK)
  Genomics England, 65
  Human Fertilisation and Embryology Authority (HFEA), 117
  Royal Society, 111*n*5
United Nations (UN)
  Leadership Council of the Sustainable Development Solutions Network, 28
  Security Council Resolution 1540, 139
United States (US, USA)
  Department of Agriculture (USDA), 117, 121, 125*n*11, 196
  Department of Health and Human Services, 16
  drug therapy model, 188, 189
  Federal Bureau of Investigation), 134
  Food and Drugs Administration (FDA), 187
  gene editing developments, marketable products derived from, 119
  Genetic Information Non Discrimination Act (GINA), 78
  National Academy of Medicine, 111*n*5
  National Academy of Sciences (NAS), 100, 111*n*5, 197
  National Institutes of Health (NIH), 16, 55, 171, 179, 186
  Precision Medicine Initiative, 65
  repatriation of international workers, 40
  safety governance in, 88
United States' National Research Council, 141
Universality, 66
Usability, 74
Use of data, 50–53
US Patent and Trademark Office (USPTO), 107

Vegetarian Diet, 172, 174, 176, 177
Viber, 46
Visual inspection of the cervix with acetic acid (VIA), 15
Volumetrics Diet, 173
Voluntarism, 151
Vulnerability, 7, 21, 28*n*9, 56, 147, 148, 158, 162
  allocational, 156
  cognitive, 156
  deferential, 156
  ethical principle of, 155–157
  medical, 156
  situational, 156
  social, 156

Weapon, defined, 137
Web of Science, 170
Weight Watchers Diet, 172
Wellcome Trust, 3, 65
Western medicine, 186
WhatsApp, 46, 52
Whole food plant-based diet (WFPB), 189, 191–196
Wholism, 194, 195
Working Group of Indigenous Minorities in Southern Africa (WIMSA), 17
World Health Organization (WHO), 5, 7, 16, 36, 148
World Wide Web, 46

Zayner, Josiah, 121
ZFN-3 technique, 101, 106
Zhang, Feng, 119
Zinc Finger Nucleases (ZFNs), 116